STUDENT UNIT

AS Biology
UNIT 3

OCR

Module 2803:
Transport/Experimental Skills 1

Richard Fosbery

AS Biology

Philip Allan Updates
Market Place
Deddington
Oxfordshire
OX15 0SE

tel: 01869 338652
fax: 01869 337590
e-mail: sales@philipallan.co.uk
www.philipallan.co.uk

© Philip Allan Updates 2004

ISBN 0 86003 672 3

All rights reserved; no part of this publication may be reproduced, stored in a retrieval system, or transmitted, in any form or by any means, electronic, mechanical, photocopying, recording or otherwise without either the prior written permission of Philip Allan Updates or a licence permitting restricted copying in the United Kingdom issued by the Copyright Licensing Agency Ltd, 90 Tottenham Court Road, London W1P 9HE.

This guide has been written specifically to support students preparing for the OCR AS Biology Unit 3 examination. The content has been neither approved nor endorsed by OCR and remains the sole responsibility of the author.

Printed by Information Press, Eynsham, Oxford

Contents

Introduction
About this guide 4
AS biology 5
The module test 5
Experimental and investigative skills 7

■ ■ ■

Content Guidance
About this section 10
Transport
Transport systems 11
The mammalian transport system 13
The mammalian heart 26
Transport in multicellular plants 30
Experimental and investigative skills
Skill P: planning 38
Skill I: implementing 45
Skill A: analysing 47
Skill E: evaluating 51
Microscopy in the practical test 54

■ ■ ■

Questions and Answers
About this section 58
Q1 Blood, tissue fluid and lymph 59
Q2 Transport of oxygen 62
Q3 Heart structure and the cardiac cycle 65
Q4 Transpiration and the transport of water 69
Q5 Translocation 72

AS Biology

Introduction

About this guide

This unit guide is the third in a series of three, which cover the OCR AS specification in biology. It is intended to help you prepare for Unit 3, which examines the content of **Module 2803/01: Transport** and assesses your practical and investigative skills in **Module 2803/02: Coursework 1** and **Module 2803/03: Practical Examination 1**. It is divided into three sections:

- **Introduction** — this gives advice on how to use the guide to help your learning and revision and on how to prepare for the examination.
- **Content Guidance** — here you will find key facts, key concepts and links with other parts of the AS/A2 biology course; you should find the links useful in your practical work and in preparing for the other units. This section also gives advice on the skills needed for your coursework and for the practical examination. It is likely that your school or college will have decided which of these modules you are going to take.
- **Questions and Answers** — here there are five questions on topics that make up Module 2803/01, together with answers written by two candidates and examiner's comments.

This is not just a revision aid. This is a guide to the whole unit and you can use it throughout the 2 years of your course if you decide to go on to A2.

The Content Guidance section will help you to:
- organise your notes and to check that you have highlighted the important points (key facts) — little 'chunks' of knowledge that you can remember
- understand how these 'little chunks' fit into the wider picture of biology; this will help:
    - to support Modules 2801 and 2802; your knowledge of the other two modules will help you understand much of the content of this one
    - to support the A2 modules, if you decide to continue the course

In addition, this section gives you advice on:
- planning and carrying out an investigation
- analysing results
- evaluating procedures
- evaluating data

The Question and Answer section will help you to:
- check the way examiners ask questions at AS
- understand what the examiners' command terms mean
- interpret the question material, especially any data that the examiners give you
- write concisely and answer the questions that the examiners set

AS biology

The diagram shows the three units that make up the AS course.

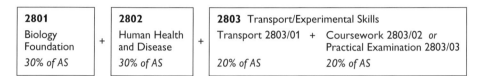

The specification outlines what you are expected to learn and do. The content of the specification for Module 2803/01 is written as **learning outcomes**; these state what you should be able to do after studying and revising each topic. Some learning outcomes are precise and cover just a small amount of factual information. Others are much broader. Do not think that any two learning outcomes will take exactly the same length of time to cover in class or during revision. It is a good idea to write a glossary to the words in the learning outcomes; the examiners will expect you to know what they mean. This guide should help you to do this.

The module test

The paper will be printed in a booklet, in which you will write all your answers. The paper will have about five questions, each divided into parts. These parts comprise several short-answer questions (no more than 4 or 5 marks each) and one question requiring an extended answer, for no more than 7 or 8 marks. In the extended-answer question, 1 mark is awarded for quality of written communication (QWC). This is used to reward spelling, grammar, punctuation, legibility, use of specialist terms and organisation of ideas. The module test has a total of 45 marks and lasts 45 minutes.

Command terms

You need to know how to respond to the various command terms used in the unit test. These are outlined below.

'Describe' and 'explain'

These do not mean the same thing! 'Describe' means give a straightforward account. You may be asked to describe something on the paper, such as a graph. You may have to describe a structure or 'tell a story', for example by writing out the sequence of events in the cardiac cycle. 'Explain' means that you should give some *reasons* why something happens.

'Name', 'identify' and 'state'

These all require a concise answer, maybe just one word, a phrase or a sentence.

'Calculate' and 'determine'
Expect to be tested on your numeracy skills. The examiner may ask you to calculate a percentage, a percentage change or a mean value for a set of figures. 'Determine' means more than just calculate. You may be asked to explain how measurements should be taken and how a final answer is calculated.

'Outline'
This means give several different points about the topic without concentrating on one or giving a lot of detail.

'Draw', 'sketch', and 'complete'
'Draw' and 'sketch' mean draw something on the examination paper, such as a graph, drawing or diagram. 'Complete' means that there is something that you need to finish, like a table, diagram or graph.

'Credit will be given for using the data'
You should look at the figure or table and use some of the information in your answer. You may do this by quoting figures (with their units) or by identifying a trend and using the information to illustrate the trend you have described.

'Differences'
It is likely that you will be asked to say how 'A differs from B'. The examiners will assume that anything you write will be something about A that is not the same as for B. Sometimes there will be a table to complete to show differences.

Prepare yourself

Make sure that you have two or more blue or black pens, a couple of sharp pencils (preferably HB), a ruler, an eraser, a pencil sharpener, a watch and a calculator.

When told to start the paper, look through all the questions. Find the end of the last question (it may be on the back page — don't miss it). Find and read the question that requires an extended answer. Some points may come to mind immediately — write them down before you forget.

There is no need to start by answering question 1, but the examiner will have set something quite straightforward to help calm your nerves. Look carefully at the number of marks available for each part question. Do not write a lengthy answer if there are only 1 or 2 marks available. If you want to change an answer, then cross it out and rewrite the answer clearly. If you write an answer or continue an answer somewhere other than on the allotted lines, then indicate clearly where this is.

When you reach the question that requires an extended answer:
- plan what you intend to write and make sure it is in a logical sequence
- do not write out the question
- keep to the point — you do not need an introduction or a summary
- use bullet points if they help your answer

- use diagrams if they help your answer — remember to label and annotate them
- pay careful attention to spelling, punctuation and grammar

Time yourself. Work out where you expect to be after about 20 minutes. Leave yourself at least 5 minutes to check your paper to make sure you have attempted all the questions and have left nothing out. The best way to do this is to check the mark allocation — have you offered something for each mark?

Experimental and investigative skills

You are tested on four skills:
- Skill P — planning
- Skill I — implementing
- Skill A — analysing
- Skill E — evaluating

The OCR specification has a set of descriptors for each of these skills areas. You can find them in the specification and you should refer to them when you are doing your coursework and when you are preparing for the practical examination. The section on coursework and the practical examination (pages 38–55) should help you understand how these descriptors relate to your practical work.

Coursework

If you take the coursework module, then you have to submit a mark for each of the four skills. The maximum marks are as follows:
Skill P — 8
Skill I — 7
Skill A — 8
Skill E — 7
Total = 30

Your total mark for the coursework is doubled to give a mark out of 60.

It is usual for candidates to undertake one piece of coursework from which they can gain all four marks. If you do this, then you will:
- plan an investigation (Skill P)
- implement your plan (Skill I)
- analyse the results you obtain and draw some conclusions (Skill A)
- evaluate your method and the data you collect (Skill E)

Alternatively, your teacher may set a planning exercise that you do not carry out in full. Instead, you carry out some preliminary work and write a plan but do not implement it to get a full set of results. Your teacher may then give you an investigation that he/she has planned. You will then implement the plan, record results and then carry out an analysis and evaluation. This means that your coursework marks will come from two different activities. It is possible for the marks to come from four different activities — one for each skill — but this is unlikely. One word of warning — the mark for any one skill (P, I, A and E) must come from one piece of work. For example, you cannot get your mark for Skill P by taking some aspects from one plan and the rest from a different plan.

Your teacher will mark your coursework. The marks will be submitted to OCR. A moderator appointed by OCR will then check your teacher's marking. Marks may be adjusted at this stage, so do not assume that the score from your teacher is your final mark.

Practical examination

There are two components to the practical examination: the **planning exercise** and the **practical test**.

The planning exercise

The planning exercise is similar to the planning component of the coursework and the advice given on pages 38–45 is identical. Your teacher does not choose the topic for the planning exercise: an examiner sets the investigation and all candidates taking the examination have the same task. Your plan will be collected by your teacher just before the date of the examination and sent to an examiner who will mark it.

The practical test

The practical test has two questions. Question 1 is usually a practical task that you complete in about an hour. You follow instructions and take results (Skill I) and then answer questions that will involve analysing results and drawing conclusions (Skill A) and evaluating the practical task and the data you collect (Skill E). Question 2 is usually based on one or two microscope slides and/or photographs. You will need to look carefully at the image(s), make some simple drawings and write some observations on what you see. In both questions you will be expected to use your knowledge of AS biology from all three modules. The practical task (usually Question 1) tends to be set on the same topic area as the planning exercise. This means that the research you do for the plan will be useful for answering questions in the test. You may have your plan to hand during the practical test if you wish to refer to it.

Scientific knowledge and understanding

The practical examination is set on the AS course. If you are entering coursework, then it is important to realise that you should undertake an investigation into a topic that is in the AS specification. You should use the scientific knowledge and understanding (SKU) that you have been taught. AS textbooks are a good guide to the detail and depth of knowledge that you should use. SKU is important in planning and analysing.

Content Guidance

This section is a guide to the content of Module 2803.

The first part deals with Module 2803/01: Transport. 'Key facts' are presented as easy-to-remember 'chunks' of knowledge. This section will also help you to understand the links with other parts of the AS and A2 course and with practical work.

There are two other modules to consider:
- 2803/02 Coursework 1
- 2803/03 Practical Examination 1

Your teacher will decide which of these two you will take. It is worth pointing out that there is no specific content for these two modules. There are no learning outcomes, as for the other three AS modules. The content of the coursework and the content of the practical examination may come from 2801 (Biology Foundation), 2802 (Human Health and Disease) and 2803/01 (Transport). You should expect to be taught about the experimental and investigative skills that you need for your coursework or for the practical examination.

Units

Various units are used in this module. This is just to remind you.

Volume: cm^3 and dm^3; $1000\,cm^3 = 1\,dm^3$

You will often find ml (millilitre) on glassware and in books. Examination papers, however, use cm^3 (cubic centimetre or 'centimetre cubed') and dm^3 (cubic decimetre or 'decimetre cubed'). $1\,cm^3 = 1\,ml$; $1\,dm^3 = 1$ litre (1 l or 1 L).

Length: nm, µm, mm, m and km; 1000 nm (nanometres) = 1 µm (micrometre); 1000 µm = 1 mm (millimetre); 1000 mm = 1 m (metre); 1000 m = 1 km (kilometre)

In this module, you may be expected to find the measurements or magnifications of cells, such as red and white blood cells.

Pressure: Pa (pascals) and kPa (kilopascals); 1000 Pa = 1 kPa

The medical profession uses mmHg (millimetres of mercury) to measure blood pressure. Examination papers use kilopascals. 'Normal' blood pressure is often given as '120 over 80' or 120 mmHg (systolic) and 80 mmHg (diastolic). These are equivalent to 15.8 kPa and 10.5 kPa (see page 29). Kilopascals are used in the Transport module for partial pressures of oxygen and carbon dioxide and also for water potentials (see pages 30–31).

Transport

Transport systems

Key concepts you must understand

Humans are fairly large multicellular organisms. Although there are many organisms much larger than us, there are a vast number that are smaller. Size is important when it comes to exchanging substances, especially oxygen and carbon dioxide, with the surroundings and then moving them around the body. Figure 1 shows *Amoeba*, which is a small organism that has a body consisting of one mass of cytoplasm. It is an example of a unicellular (one-celled) organism. *Amoeba* is non-photosynthetic and gains its energy by eating smaller organisms, such as bacteria. It lives in freshwater. Figure 1 shows the exchange of gases that occurs between *Amoeba* and its surroundings. The cell surface membrane serves as the site of gaseous exchange. The surface area is large enough to provide sufficient oxygen for respiration and to remove the waste carbon dioxide.

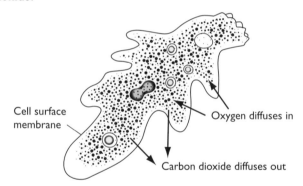

Figure 1 *Amoeba is a unicellular organism that has a large surface area to volume ratio. It uses its body surface for gaseous exchange.*

Larger, multicellular animals, such as fish, insects and squid, do not have sufficient body surface to act as the site of gaseous exchange. There just is not enough surface to absorb the amount of oxygen required. This is why animals have lungs or gills with large surface areas as sites of gaseous exchange. There are specialised surfaces for exchange elsewhere in the body. For example, the lining of the gut has a large surface area for the absorption of digested food.

Substances such as oxygen, carbon dioxide and absorbed food have to be moved around the body. In unicellular organisms, such as *Amoeba*, substances can pass by diffusion or be carried in cytoplasm as it flows within the organism. This does not work in a large organism because distances are too great. Oxygen cannot be supplied fast enough by diffusion from the lungs to cells deep in the body; a transport system is needed. In this module you study three systems — blood, xylem and phloem. All three are examples of **mass flow** — the movement of a fluid through a system of tubes in one direction.

Table 1 compares the transport mechanisms in flowering plants and mammals.

Feature	Mammals	Flowering plants
Transport system	Circulatory system — heart + blood vessels + blood	Xylem and phloem
Gas exchange surface	Alveoli in the lungs	All cell surfaces that are in contact with the air — e.g. palisade and spongy mesophyll cells in leaves
Transport of oxygen	Oxygen in combination with haemoglobin	Oxygen and carbon dioxide diffuse through air spaces between cells
Transport of carbon dioxide	Carbon dioxide in blood plasma as HCO_3^- and in combination with haemoglobin	
Transport of carbohydrate	Glucose in solution in blood plasma	Sucrose in solution in phloem sap
Transport of water	Most of the blood plasma is water	In the xylem sap
Force to move fluids	Hydrostatic pressure generated by contraction of the heart	Xylem — transpiration pull Phloem — active pumping of sugars into the phloem + hydrostatic pressure

Table 1

It is difficult to calculate the surface area for animals and plants, but you need to know about how the **ratio** of surface area to volume changes as organisms increase in size. It helps to use cubes of different sizes to understand this principle. Table 2 shows what happens to the surface area to volume ratio as a cube increases in size.

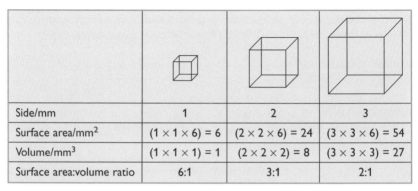

Side/mm	1	2	3
Surface area/mm²	(1 × 1 × 6) = 6	(2 × 2 × 6) = 24	(3 × 3 × 6) = 54
Volume/mm³	(1 × 1 × 1) = 1	(2 × 2 × 2) = 8	(3 × 3 × 3) = 27
Surface area:volume ratio	6:1	3:1	2:1

Table 2

- Small organisms have a *large* surface area to volume ratio.
- Large organisms have a *small* surface area to volume ratio.

Small organisms use their body surface for gaseous exchange but larger ones have specialised surfaces for exchange.

Links Most examination papers in biology have questions that involve a calculation. You may have to calculate surface area to volume ratios. Remember to divide the surface area by the volume. If you do this on your calculator, you will see a number that represents how much surface area (in units of area, e.g. mm^2) there is for every unit of volume, e.g. for every 1 mm^3. Write this down as a ratio, for example A:1 where A is the number on your calculator. You may not get a whole number. If this is the case, round up or down to one decimal place. It is acceptable to write a SA:V ratio as something like 1.5:1.

You will meet the principle of *large surface area* many times in biology. You have probably come across it in Module 2801, where cell surface membranes provide large surface areas for exchange of substances. Animal cells have microvilli to increase their surface area for exchange.

Some large organisms, such as jellyfish, make use of their surface for gaseous exchange. They have a large SA:V ratio because their bodies are extended into tentacles, and they do not have any deep tissue — their bodies are made of two layers of cells.

The mammalian transport system

Key concepts you must understand

There are three important components of this transport system:
- the blood — red blood cells, white blood cells, platelets and plasma
- blood vessels — arteries, arterioles, capillaries, venules, veins
- the heart — the pump for circulating blood through the vessels

Figure 2 shows a simple view of the mammalian circulation.

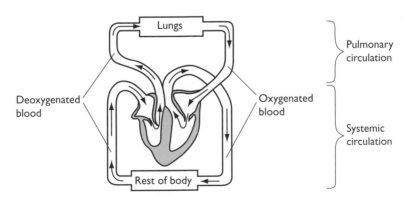

Figure 2 The mammalian circulation. Follow the pathway from the lungs and back to see why it is called a double circulation.

There are two important features that you can see.
- The blood flows inside blood vessels — it is a **closed circulation** (perhaps an *enclosed* circulation is a way to remember this).
- The blood flows through the heart twice to complete one circuit of the body — it is a **double circulation**.

The circuit from the heart to the lungs and back is the **pulmonary circulation**. The circuit from the heart to the rest of the body and back is the **systemic circulation**.

What, you might ask, are open and single circulations? Insects have an open circulation. There are some blood vessels but the blood flows out of these to bathe the tissues directly, rather than travelling in capillaries. Fish have a closed circulation, in which the blood passes from the heart to the gills and then immediately to the rest of the body. It passes through the heart once during a complete circulation of the body. Fish have a single circulation.

The important function of the blood is to transport gases (oxygen and carbon dioxide), nutrients, hormones and waste products, such as urea. During each circuit of the body, blood flows through capillaries in the lungs and in other organs, such as the stomach, liver and kidney. Figure 3 shows the pathway taken by blood as it flows through a capillary bed in an organ. Capillaries are the **exchange vessels** of the circulatory system. It is here that substances pass into and out of the blood.

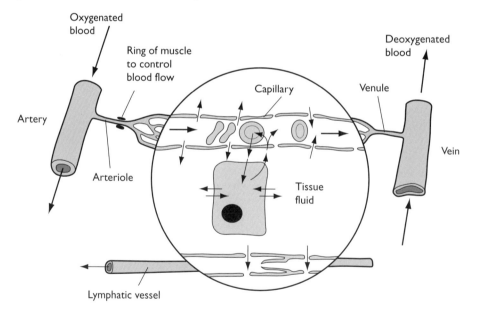

Figure 3 Blood is supplied to an organ by an artery. It then flows through arterioles and then capillaries where exchanges occur between blood and tissue fluid. Blood drains through venules and then veins to return to the heart. Capillaries are very small, which is why the central area here is shown magnified.

Key facts you must know

Figure 4 shows cross-sections of an artery and a vein.

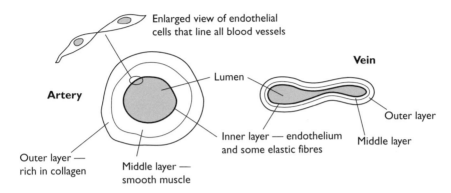

Figure 4 Cross-sections of an artery and a vein

Table 3 compares the structure and functions of arteries and veins.

Feature	Artery	Vein
Width of wall	Thick	Thin
Components of wall	Smooth muscle, elastic fibres, collagen	As artery, but less of each tissue
Semi-lunar valves	✗	✓ (to prevent backflow)
Blood pressure	High	Low
Direction of blood flow	Heart to tissues	Tissues to heart
Function	Elastic fibres recoil to maintain high pressure in arteries to overcome resistance of the circulatory system	Return blood to the heart — assisted by squeezing action of surrounding muscles, which help to push blood towards heart

Table 3

As blood flows around the body its pressure changes. There is a need for blood pressure — without it the blood would not flow through the blood vessels. The vessels present a resistance to blood flow and the contraction of the heart raises the pressure of the blood, forcing it through the circulation. There is a high pressure in the arteries so that blood is delivered efficiently to organs. When it gets there, however, it would burst the tiny capillaries if the pressure were not 'damped down'. Arterioles are the blood vessels that achieve this 'damping down' effect. They do this by contracting. An arteriole can contract to reduce the size of its lumen so that blood flow to the capillaries decreases. This allows blood to be diverted elsewhere, for example from skin to muscles. There is only enough blood to fill 25% of the capillaries at any one time, so arterioles play an important function in controlling the flow of blood to tissues, rather like taps turning the flow of water on and off. Changes in blood pressure in the systemic circuit are shown in Figure 5.

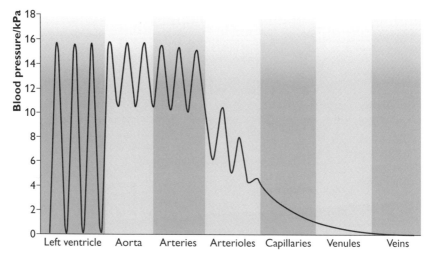

Figure 5 The changes in blood pressure at different places in the systemic circulation (not drawn to scale). Arteries and veins may be metres in length (think of a blue whale or a giraffe); arterioles, capillaries and venules are only a few millimetres in length.

The graph shows the pressure changes in three successive heart beats and then the effect of these in the arteries and other vessels. Notice how there are three peaks and troughs in the left ventricle and then see how these are repeated in the aorta and the arteries. The rise and fall in blood pressure is greatest in the left ventricle. This is then reduced in the aorta and main arteries and becomes much smaller in the arterioles, where the blood pressure decreases considerably to protect the capillaries. Blood pressure decreases further in the capillaries because of friction between the blood and the walls. Blood pressure is lowest in the venules and veins. From this graph of blood pressure you can see why arteries have thick, muscular and elastic walls (to withstand high blood pressure), arterioles have muscular walls (to damp down blood pressure) and veins have thin walls (as the blood has a low pressure). Capillaries have thin walls as they are exchange vessels (see Figures 6 and 7).

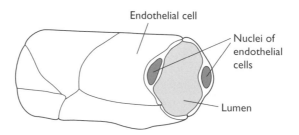

Figure 6 A capillary lined by endothelial cells

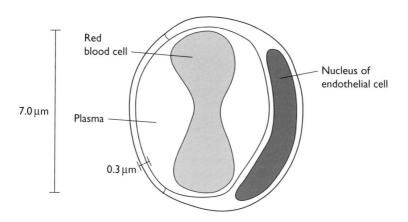

Figure 7 Cross-section of a capillary. Notice that red blood cells just fit inside the vessel and that the endothelial cells are thin to help the exchange of substances with the blood.

Exchanges that occur as blood flows through capillaries include:
- oxygen diffusing out of the blood into tissue fluid
- carbon dioxide diffusing from tissue fluid into the blood
- water and dissolved substances, such as glucose and amino acids, being forced out by the pressure of the blood

Water and substances that are forced out by high pressure are **filtered** from the blood. As blood flows through the capillaries its pressure decreases, which makes it possible for water to drain back into the blood plasma by osmosis. This is because the blood contains solutes, such as albumen, which give the blood plasma a lower water potential than the tissue fluid. Albumen is a large protein molecule that rarely leaves the blood.

Feature	Blood	Tissue fluid	Lymph
Where found	In blood vessels	Surrounding cells	In lymphatic vessels
Components:			
• Red blood cells	✓	✗	✗
• White blood cells	✓	✓ (some)	✓ (some)
• Fats	✓ (as lipoproteins)	✓	✓ (especially after a meal)
• Glucose	✓	✓	✓ (very little)
• Proteins	✓	✓ (some)	✓ (some, e.g. antibodies)
Functions	Transport	Bathes cells — all exchanges between blood and cells occur through tissue fluid	Drains excess tissue fluid, preventing a build up that would lead to oedema

Table 4 The composition and functions of three body fluids

Not all the tissue fluid is drained this way. Within the tissues are small, blind-ended tubes called lymphatic vessels, which act as a 'drainage system'. Tissue fluid flows into the lymphatic vessels and then flows quite slowly towards large lymphatic vessels that empty into the blood near the heart. The fluid inside lymphatic vessels is called lymph. It is similar in composition to tissue fluid. At intervals along the lymphatic vessels are lymph nodes. These contain lymphocytes, some of which flow into the blood via the lymph. Lymph also drains fat from the small intestine so that after a meal it often appears as a white suspension.

Blood cells

You should be able to recognise the types of blood cells shown in Figure 8, both in photographs and through the microscope.

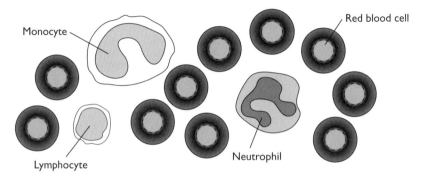

Figure 8 A drawing of a blood smear, as seen through a light microscope

Table 5 compares the structure and function of the main types of blood cell. The most common type of phagocyte in the blood is the neutrophil. About 70% of all white blood cells are neutrophils. Monocytes are transported in the blood from the bone marrow, where they are produced, to tissues, such as lymph nodes, where they become macrophages.

Feature	Red blood cells	Phagocytes	Lymphocytes
Diameter/μm	7	9	4 to 6
Nucleus	✗ (lost during development in bone marrow)	✓ (lobed)	✓ (large — fills most of the cell)
Organelles	None — cytoplasm is full of haemoglobin molecules	Present — e.g. mitochondria; lysosomes containing enzymes to break down bacteria and other pathogens	Present — especially rough endoplasmic reticulum and Golgi body in activated B cells (also known as plasma cells)

Table 5

Links You need to know about two types of lymphocyte in Module 2802. These are B and T lymphocytes, which have different functions during an immune response. Activated B lymphocytes divide to become plasma cells, which secrete antibody molecules. Plasma cells are full of rough endoplasmic reticulum for the production of protein and have a Golgi body for the packaging and release of antibody molecules. Remember that antibodies are made of protein. T-helper cells stimulate B lymphocytes to divide and T-killer cells search out and kill cells that are infected with viruses.

Gaseous exchange in the lungs and transport of gases

Gaseous exchange occurs as blood flows around the alveoli in the lungs. Figure 9 shows how the alveoli are adapted for this.

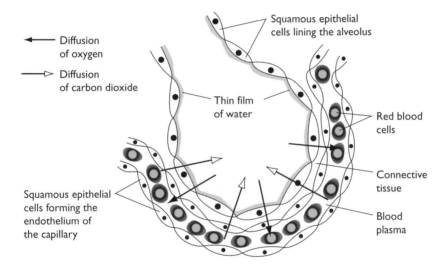

Figure 9 An alveolus — the site of gaseous exchange

A molecule of oxygen from the air in the alveolus dissolves in the water film. It then diffuses through the following:
- squamous epithelium
- connective tissue
- capillary endothelium
- blood plasma
- membrane of a red blood cell

Carbon dioxide from the blood diffuses in the reverse direction.

Key facts you must know

Cells respire. To respire aerobically they need a supply of oxygen and they need the waste carbon dioxide removed. Blood is adapted to achieve this transport. If we only had a watery fluid to transport oxygen and carbon dioxide, we would not be able to carry these gases efficiently. Oxygen is not very soluble in water. The volume of oxygen that will dissolve in water is about $0.3\,cm^3$ of oxygen per $100\,cm^3$. Blood can carry $20\,cm^3$ of oxygen per $100\,cm^3$. Carbon dioxide is much more soluble in water than oxygen is, and about $2.6\,cm^3$ per $100\,cm^3$ could be transported in solution. However, blood can carry $50\text{–}60\,cm^3$ of carbon dioxide per $100\,cm^3$. How is this possible?

At this point, it is a good idea to review the structure of haemoglobin (Unit 1).

Haemoglobin (Hb for short) transports almost all the oxygen in the blood and also some of the carbon dioxide. Each red blood cell has about 280 million molecules of haemoglobin. Each haemoglobin molecule can combine with four molecules of oxygen to form oxyhaemoglobin:

$$Hb + 4O_2 \rightleftharpoons HbO_8$$

When carbon dioxide dissolves in water, quite a lot of it reacts to form carbonic acid, which dissociates ('breaks up') to form hydrogen ions and hydrogencarbonate ions:

$$CO_2 + H_2O \rightleftharpoons H_2CO_3 \rightleftharpoons H^+ + HCO_3^-$$

carbon dioxide + water ⇌ carbonic acid ⇌ hydrogen ions + hydrogencarbonate ions

Blood can carry much more carbon dioxide than the volume that dissolves in water because there is a fast-acting enzyme inside red blood cells. This enzyme is **carbonic anhydrase**. It catalyses the formation of carbonic acid, which immediately dissociates to form hydrogen ions and hydrogencarbonate ions. Large quantities of hydrogencarbonate ions are carried in the blood plasma in association with sodium ions. Some carbon dioxide also attaches to the $-NH_2$ groups (amino groups) at the end of the polypeptides to form carbamino-haemoglobin, and some remains in solution (as CO_2) in the plasma.

Key concepts you must understand

Transport of oxygen

In the lungs, deoxygenated blood flows very close to alveolar air. The air in the alveoli is rich in oxygen. This 'richness' is expressed as its partial pressure (abbreviated to pO_2). Partial pressure is that part of the air pressure that oxygen exerts. About 13–14% of the air inside the alveoli is oxygen. The total air pressure is about $100\,kPa$, so the partial pressure of oxygen in the alveoli is about $13\text{–}14\,kPa$. (See page 10 if you are unsure about pressure units.) Deoxygenated blood flowing into the lungs has a low concentration of oxygen. Alveolar air is rich in oxygen, so a concentration gradient exists between the air and the blood and oxygen diffuses from the air in the alveoli into the blood.

The oxygenated blood leaving the lungs carries almost the full amount of oxygen it can possibly carry. Tissues such as muscle tissues and those in the gut, liver and kidney use oxygen in respiration. The concentration of oxygen in these tissues is low, equivalent to a partial pressure of about $5.0\,kPa$. This means that oxygen diffuses from the blood into the tissues down a concentration gradient. Blood loses about 30% of the oxygen it carries as it flows through the tissues when you are at rest, that is, not doing any exercise.

To find out how much oxygen is transported by haemoglobin, small samples of blood are exposed to gas mixtures with different concentrations of oxygen. The volume of oxygen absorbed by the haemoglobin in each sample of blood is determined and expressed as a percentage of the maximum volume that haemoglobin absorbs. The results are shown as an oxygen haemoglobin dissociation curve in Figure 10. The concentration of oxygen is shown as the partial pressure of oxygen in the gas mixture.

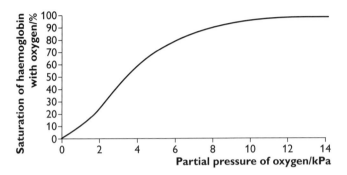

Figure 10 The results of an investigation into the effects of different air mixtures on the saturation of haemoglobin with oxygen

The line of best fit drawn on the graph is known as the oxygen haemoglobin dissociation curve. The sigmoid shape (or S shape) of the line is important. Remember that the graph shows the results of an experiment on blood done in the laboratory. How does this relate to the actual situation in the mammalian body? If we take some figures from the graph, we should be able to see how to use this information. Table 6 shows the saturation of haemoglobin with oxygen at different partial pressures. The partial pressures chosen correspond with those in different parts of the circulation.

Site in the body	Partial pressure of oxygen/kPa	Saturation of haemoglobin with oxygen/%
Lungs	13.0	98
Tissues, including muscles at rest	5.0	70
Muscles during strenuous exercise	3.0	43
Muscles at exhaustion	2.0	24
Placenta	4.0	60

Table 6

Notice from Figure 10 and Table 6 that haemoglobin is:
- nearly fully saturated at the partial pressure of oxygen in the lungs
- about 70% saturated at the partial pressure in the tissues
- about 45% saturated in areas with very low partial pressures of oxygen, as in actively respiring muscle

As blood flows through capillaries, oxyhaemoglobin releases oxygen *in response to the low concentration of oxygen in the tissues*. Haemoglobin has a high affinity for oxygen at high partial pressures and a low affinity at low partial pressures.

Some animals live at high altitude, where the air is thinner than at sea level and there is a lower partial pressure of oxygen. The haemoglobin of these animals has a higher affinity for oxygen than the haemoglobin of those that live at sea level.

Fetal mammals have the same problem as mammals that live at high altitude. Gaseous exchange between maternal blood and fetal blood occurs across the placenta. In order to be nearly fully saturated with oxygen, fetal haemoglobin must have a higher affinity for oxygen than adult haemoglobin. When experimenters investigated fetal blood using the same method as used to give the results shown in Figure 10, they found when they plotted the dissociation curve that it was *to the left* of the curve for adult haemoglobin. The partial pressure of oxygen at the placenta is about 4.0 kPa. At this partial pressure, adult blood is about 60% saturated, which means that oxyhaemoglobin will give up its oxygen to the surrounding tissues in the placenta. Fetal haemoglobin is about 80% saturated at this partial pressure, so it absorbs much of the oxygen that is released by the mother's blood. Tissues in the placenta absorb the rest.

Transport of carbon dioxide

Figure 11 shows the events that occur as carbon dioxide diffuses into the blood in tissues.

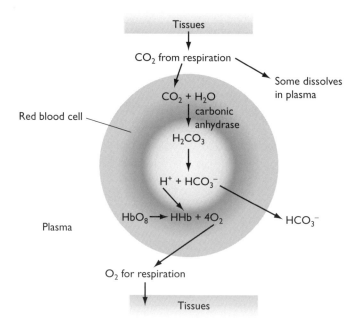

Figure 11 *Most of the carbon dioxide diffuses into red blood cells and is converted to hydrogencarbonate ions by the action of carbonic anhydrase. Some carbon dioxide dissolves in the plasma.*

When the changes shown in Figure 11 occur, hydrogen ions (H⁺) are released. These would reduce the pH of blood cells if left unchecked. Haemoglobin binds these hydrogen ions to become haemoglobinic acid (HHb). This prevents a decrease in pH. Haemoglobin acts as a **buffer** by doing this. It buffers the change in pH, stopping it from decreasing.

When the blood reaches the lungs, these changes go into reverse, as shown in Figure 12.

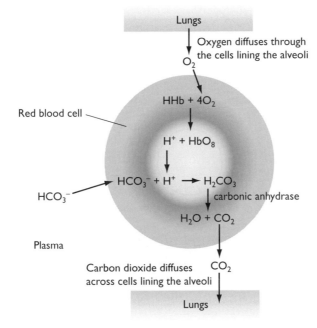

Figure 12 These events occur as blood flows through capillaries in the alveoli, so that carbon dioxide diffuses into alveolar air

The Bohr effect

When carbon dioxide diffuses into the blood, it stimulates haemoglobin to release even more oxygen than it would if it only responded to the low concentration of oxygen. This can be explained by the fact that when haemoglobin accepts hydrogen ions to become haemoglobinic acid (HHb) it stimulates the molecule to give up oxygen (see Figure 11). When the experiment with gas mixtures was repeated using carbon dioxide, it was discovered that carbon dioxide *decreased* the affinity of haemoglobin for oxygen. Figure 13 shows the effect when the results are plotted on a graph.

This shift of the curve to the right is known as the Bohr effect, after the Danish scientist who discovered it. The best way to understand the Bohr effect is to take some figures from the graph. Table 7 shows the effect of increasing the partial pressure of carbon dioxide on the saturation of haemoglobin. The figures are taken for the *same* partial pressure of oxygen (pO_2 = 3.0 kPa, which corresponds with that in the tissues when

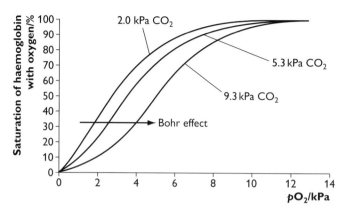

Figure 13 The Bohr effect. When carbon dioxide was added to the gas mixtures, haemoglobin became less saturated with oxygen.

they are respiring actively). Notice that the haemoglobin is much less saturated with oxygen when the pCO_2 is 9.3 kPa. As blood flows through the capillaries it gives up even more oxygen than it would have done if there were less carbon dioxide present.

Partial pressure of carbon dioxide/kPa	% saturation of haemoglobin with oxygen at pO_2 of 3.0
2.0	55
5.3	50
9.3	20

Table 7

The effect of altitude on the blood

Air at high altitudes has a lower partial pressure of oxygen than air at sea level. Mammals adapted to live at high altitudes have haemoglobin with a higher affinity for oxygen that those at sea level. If we travel to high altitude, we cannot make a different type of haemoglobin. To compensate, we make more red blood cells. Our haemoglobin is still not fully saturated when it leaves the lungs, but the presence of extra blood cells means that the blood carries about the same volume of oxygen as it did at sea level. This acclimatisation to altitude takes several days. You can see from Figure 14 the effect on the blood of living at altitude.

Links You need to know about the mammalian transport system when studying Module 2802. To understand the sections on gaseous exchange and exercise in Module 2802, you need to know about the transport of gases by the blood. During exercise, the blood system cannot supply oxygen to the tissues fast enough and an oxygen deficit develops. When exercise finishes, the haemoglobin is often not fully saturated and you breathe heavily for a while to reoxygenate the haemoglobin and respire the lactate produced when muscle tissue respired anaerobically. During recovery from exercise, you are repaying an oxygen debt.

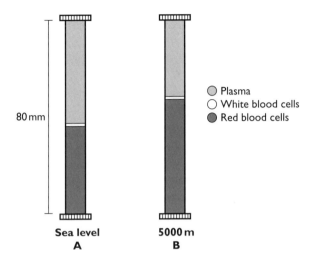

Figure 14 Tube A contains blood from a person who lives at sea level. Tube B contains blood from a person who has lived at high altitude (5000 metres) all his/her life. In both cases, a small sample of blood was spun in a centrifuge to determine the percentage of blood volume composed of red blood cells. Notice that there is a higher volume of red blood cells in B.

If fetal haemoglobin has a higher affinity for oxygen than adult haemoglobin, why don't we keep it throughout life? Unfortunately, because it has a higher affinity, it does not release its oxygen so well in respiring tissues. After birth, tissues need a much greater supply of oxygen and this is provided by adult haemoglobin — a switch occurs over the first few years of life. Sometimes this switch does not happen, with the result that the tissues are starved of oxygen. This is what happens in people who have thalassaemia, a genetic disease that occurs most frequently in people from countries around the Mediterranean. You could use thalassaemia as an example of an inherited disease when you study the categories of disease in Module 2802.

Blood cells are good examples of animal cells to use in Module 2801. Red blood cells are unusual because they have no membrane-bound organelles, such as mitochondria and endoplasmic reticulum. They have no nucleus. Does this make them prokaryotes? Certainly not! As red cells develop in bone marrow they lose their nuclei so that they can be packed with haemoglobin. So they are eukaryotic cells, if peculiar ones. In Module 2802, you have to know how phagocytes and lymphocytes function, so knowing about their structure will help you to appreciate the differences in the way they function.

In the practical examination, you may be given a microscope slide of a blood smear and be expected to draw the different types of blood cell. Make sure you have seen some of these and also some good photographs, so that you can recognise the cells. Red blood cells are easy, but it takes a while to see the difference between lymphocytes and phagocytes. Look carefully for the lobed nucleus of the neutrophil — see Figure 8. You also need to recognise arteries and veins in photographs and slides and to be able to draw them under the microscope.

The mammalian heart

Key facts you must know

The heart is a muscular pump. It is made of cardiac muscle and is described as myogenic, that is, it stimulates itself to beat. There are various features of the heart that you should know.

- There are four chambers — two atria and two ventricles.
- There are two pumps working in series — the right side of the heart pumps deoxygenated blood to the lungs in the pulmonary circulation; the left side pumps oxygenated blood to the rest of the body through the systemic circulation.
- The left and right atria have thin walls as they pump blood into the ventricles, which is only a short distance.
- The left and right ventricles have thick walls as they pump blood a greater distance and against a greater resistance than the atria.
- There are valves in the heart to prevent backflow and to ensure the blood follows the correct pathway.
- The volume of blood ejected by each chamber is the same during one beat, but the volume can change from beat to beat in response to the body's demand for oxygen.
- Stroke volume is the volume of blood ejected from each ventricle during one beat. At rest it may be about 150 cm^3. Cardiac output is the volume of blood ejected from each of the ventricles during 1 minute:

    cardiac output = stroke volume × heart rate

    At rest it may be about $150 \times 70 = 10\,500$ cm^3 = 10.5 dm^3 per minute.
- The cardiac cycle describes the sequence of changes that occurs in the heart during one heart beat.

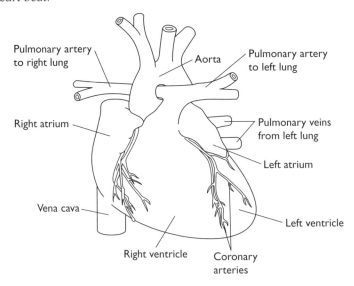

Figure 15 An external view of the heart

Drawings and diagrams of the heart usually show it viewed from the front of the body. Figure 15 shows the external structure of the heart with the major blood vessels. Note the coronary arteries that supply the heart muscle with oxygenated blood. They branch from the base of the aorta. Figure 16 shows the internal structure of the heart.

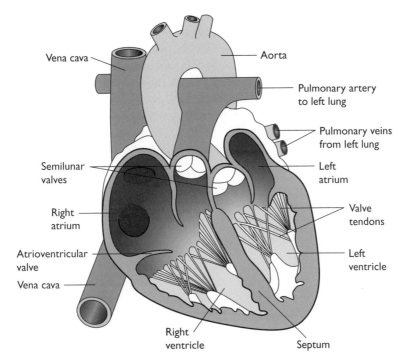

Figure 16 A vertical section through the heart

Table 8 shows the functions of the chambers of the adult human heart. The ventricles are thicker than the atria as they have to pump blood over a greater distance. The left ventricle is thicker than the right because it has to pump blood around the whole body, and the systemic circulation provides a greater resistance than the pulmonary circulation. The lungs are a spongy tissue and blood fills most of the capillaries in the lungs. There are fewer arterioles to provide resistance than in the systemic circulation.

Chamber of the heart	Receives blood from	Pumps blood to
Right atrium	Body through vena cava	Right ventricle
Right ventricle	Right atrium	Lungs through the pulmonary artery
Left atrium	Lungs through pulmonary veins	Left ventricle
Left ventricle	Left atrium	To body through the aorta

Table 8

Control of the heart

The heart beat is controlled by the sino-atrial node (SAN) in the right atrium. The SAN is a special region of muscle cells that emits pulses similar to the electrical impulses that pass along nerve cells. These travel across the muscle in the atria and cause them to contract together. The electrical impulses do not reach the ventricles directly as there is a ring of fibrous tissue between the atria and the ventricles, which prevents the impulses reaching the muscle in the ventricles. The atrio-ventricular node (AVN) is in the central septum at the junction between the atria and ventricles. The AVN delays the impulses so that they reach the ventricles after they have filled with blood from the atria. Impulses are relayed by the AVN along Purkyne tissue, which conducts to muscles at the base of the ventricles so that they contract first. This forces blood from the bottom of the ventricles upwards into the arteries. As this happens, the rest of the ventricle muscle contracts so that the ventricles empty completely.

Nerves that supply the heart alter the rate of contraction. They do not stimulate the heart to beat each time.

The cardiac cycle

The spread of impulses from the SAN starts a series of changes that comprise the cardiac cycle. Imagine that the individual drawings in Figure 17 are still pictures from a film that shows the heart beating. The small arrows in the heart show where the blood is flowing at each stage. Figure 18 shows the changes in blood pressure in the left atrium, left ventricle and in the aorta. The blood pressure is recorded by placing pressure sensors in each of these three places.

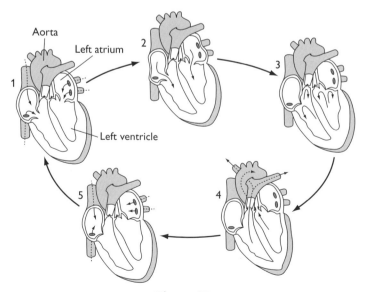

Figure 17

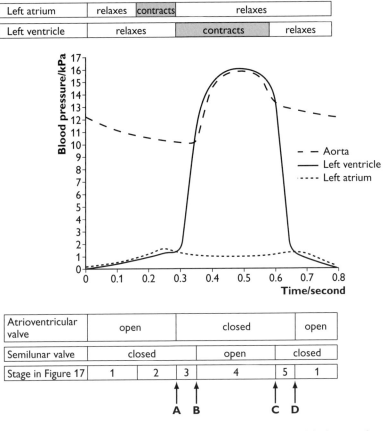

Figure 18 Contraction occurs in stages 2, 3 and 4 — this is systole.
Relaxation occurs in stages 5 and 1 — this is diastole.

Notice the following:
- The blood pressure in the atria and the ventricles falls near to 0 kPa during each cycle because there are times when there is little blood in these chambers.
- The pressure in the aorta does not fall below about 10 kPa because its wall stretches as blood surges into it from the left ventricle. The wall of the aorta then recoils to maintain the blood pressure and to keep blood flowing.
- When the pressure in the left ventricle is greater than that in the atrium, the atrioventricular valve closes to stop blood flowing back into the atria. The tendons at the base of the valve stop it 'blowing back' into the atria. This is at A in Figure 18.
- When the pressure in the ventricle is greater than that in the aorta, the semilunar valve opens and the blood flows from the ventricle to the aorta. This is at B in Figure 18.
- When the pressure in the aorta is greater than that in the ventricle, the semilunar valve fills with blood, closes off the aorta and prevents backflow. This is at C in Figure 18.
- When the pressure in the atrium is greater than the pressure in the ventricle, the atrioventricular valve opens so blood flows from the atrium into the ventricle. This is at D in Figure 18.

Now, position a ruler on Figure 18 over the vertical axis at time 0.
- Move the ruler slowly across the graph to the right and follow the changes to the blood pressure in the ventricle.
- Look at the stages (1 to 5) in Figure 17 to follow the changes to the left ventricle.
- Now move the ruler back to 0 and repeat the procedure but this time follow the changes in the atrium.
- Now repeat by moving the ruler across and following the changes in the aorta.
- Think about the opening and closing of the semilunar and atrioventricular valves. Look at the blood pressure either side of points A, B, C and D in Figure 18 and look at Figure 17 to check whether these valves are *opening* or *closing* at those points in the cycle.
- It takes 0.8 second to complete this cardiac cycle. This means that there are $\frac{60}{0.8}$ = 75 beats per minute.

To help understand this further, try to locate an animation of the cardiac cycle on the web. You can expect to be asked about the cardiac cycle graph, so it is well worth studying Figures 17 and 18 in some detail.

Links You need to apply the knowledge that you have about the heart to Module 2802. The pulse is the way we measure the heart rate. The *resting* pulse is a good way to assess **aerobic** fitness. Resting pulse is low in people who have good aerobic fitness. The blood pressure in the arteries fluctuates, as we saw in Figure 5. The highest blood pressure in the aorta and the arteries that branch from it corresponds with the contraction of the left ventricle — this is the **systolic blood pressure**. The lowest blood pressure corresponds with the time when the left ventricle is empty — this is the **diastolic blood pressure**. The blood pressure in the ventricle falls to zero, but in the arteries the elastic recoil makes sure it does not fall below about 10 kPa. Aerobic exercise makes the heart increase in size so that it can pump more blood with each beat. The stroke volume and the cardiac output increase.

Transport in multicellular plants

Key concepts you must understand

This topic is about the movement of water and assimilates, such as sucrose and amino acids. Assimilates are so called because they have been made from simple substances, such as water, carbon dioxide and ions (e.g. nitrate ions), which have been assimilated (taken up and used) by plants. We need to start with two important concepts from Module 2801 — osmosis and water potential.

Figure 19 shows the movement of water between some mesophyll cells. Cell P has a higher water potential than cells Q and R. This may be because Q and R have more solutes in them than P has, or because they are losing more water by evaporation to the air than P is. The arrows show the direction taken by water. The symbol Ψ is used to represent water potential. Remember that −300 kPa is *greater than* −400 kPa, so water moves from a higher water potential (P) to a lower water potential (Q). You should be able to state the direction water takes when given some water potentials, such as those in Figure 19.

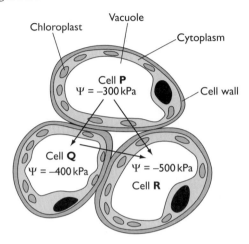

Figure 19 *The arrows show the direction of water movement between three mesophyll cells in a leaf. Cell P has the highest water potential and the water potential of cell Q is higher than that of cell R.*

There are two transport systems in plants:
- xylem for water and mineral ions
- phloem for assimilates, such as sucrose and amino acids

The contents of xylem and phloem move by mass flow. Everything within each 'tube' in these tissues moves in the same direction at the same time.

The movement of water in the xylem depends on transpiration, which is the evaporation of water from the leaves and other aerial parts of plants. The source of energy to drive transpiration comes from the sun. The plant provides a system of channels for water to flow, but does not provide energy for the movement of water in the xylem.

Movement in the phloem is translocation. Literally, translocation means from 'place to place' but it is the name given to the mass flow of assimilates dissolved in water that occurs in phloem tissue. This is driven by energy from the plant.

Key facts you must know

Figure 20 shows cross-sections of a leaf, a root and a stem of a typical flowering plant. You should be able to show on such diagrams where the xylem and phloem are situated. Figure 21 is a photograph of the central area of a root, as seen through the high power of a microscope. You should be able to identify the tissues labelled here.

AS Biology

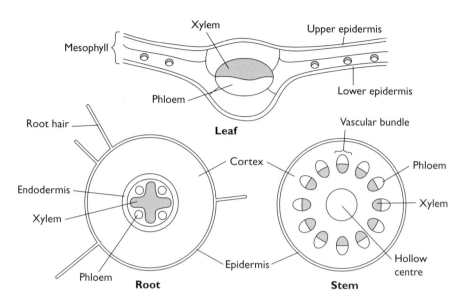

Figure 20 *The distribution of xylem and phloem in cross-sections of leaf, root and stem*

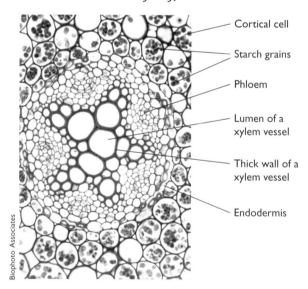

Figure 21 *This photograph shows what you should be able to see under the high power of a microscope when you look at the centre of a cross-section of a root. Notice that the xylem in this case is in the form of a five-pointed star.*

Xylem tissue consists of:
- vessel elements — dead, empty cells arranged into continuous columns called vessels
- parenchyma cells — living cells found between the vessels

Figure 22 shows details of xylem vessels and how they are adapted to transport water and provide support.

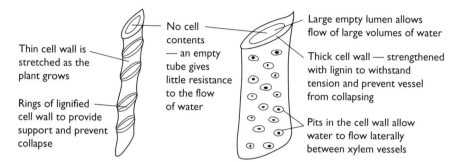

Figure 22 *Two xylem vessel elements. (a) A narrow vessel thickened with rings. (b) A wider vessel with pits to allow lateral movement of water.*

Phloem tissue consists of:
- phloem sieve tube elements — living cells arranged into continuous columns called sieve tubes
- companion cells — smaller, very active cells

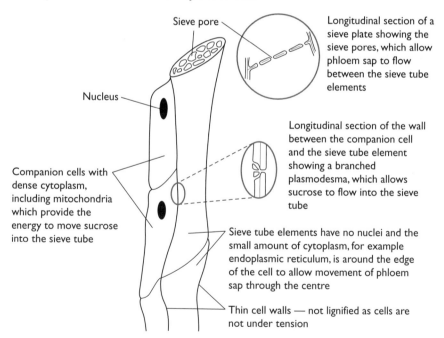

Figure 23 *Phloem sieve tubes and companion cells, showing how they are adapted for their functions. Plasmodesmata are small tubes of cytoplasm that pass through the cell wall. They are lined by membrane continuous with the cell surface membrane.*

Transport of water in plants

Most of the water absorbed by plants is lost to the atmosphere in transpiration. Water is absorbed by the roots of the plant and passes across the cortex of the root, through the endodermis and into the xylem in the central region (see Figure 21). From here it travels inside xylem vessels until it reaches the leaves, where it may enter cells and:
- be used as a raw material for photosynthesis, or
- enter the vacuole to give it turgidity and help with support, or
- pass to the cell wall and evaporate into the air spaces in the leaf

Water vapour that evaporates from cell walls may diffuse through the stomata into the atmosphere outside the leaf. As water travels across the cortex in the root and across the leaf there are two main pathways that it may follow:
- the **apoplast** pathway — along cell walls
- the **symplast** pathway — from cell to cell through the plasmodesmata

These pathways are shown in Figure 24.

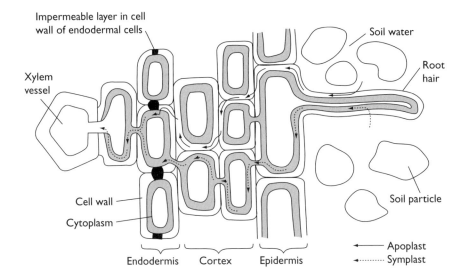

Figure 24 *The arrows show the pathways taken by water as it moves from the soil, into a root hair, across the cortex, through the endodermis and into the xylem. Plasmodesmata are shown joining the cells, but they are not as big as shown here.*

Most of the water probably follows the pathway of least resistance, which is the apoplast pathway. The endodermis is a layer of cells that have impermeable material between the cell walls. This is a barrier to the apoplast pathway, so water has to travel through cells into xylem vessels. There is a water potential gradient here, so that water travels from the cortex by osmosis across the cells and into the xylem. The function of the endodermis is probably to select ions to pump from the cortex into the xylem and then transport to the rest of the plant.

The force that draws water up through xylem vessels is transpiration pull. Water evaporates from the cell surfaces inside the leaf. This makes the air spaces inside the leaves fully saturated with water vapour. If the stomata are open, then water vapour escapes by diffusion. Transpiration is the combined effect of evaporation from the internal surfaces of leaves and the diffusion of water vapour out of the leaves. Water moves through plants because of cohesive forces between water molecules and the adhesive forces between water and cell walls. This is known as cohesion-tension. Various factors influence the rate at which transpiration occurs:

- Temperature — on hot days, the rate of evaporation increases and the air holds more water. Increasing temperature tends to increase the rate of transpiration.
- Humidity — on very humid days, the atmosphere may hold as much water as the air inside the leaves. This means that there is little or no gradient for the diffusion of water vapour. Increasing humidity tends to decrease the rate of transpiration.
- Wind speed — on windy days, water vapour molecules are blown away from the leaf surface as soon as they pass through the stomata. Increasing air speed tends to increase the rate of transpiration.
- Light intensity — in most plants, stomata open during daylight hours to obtain carbon dioxide for photosynthesis. When stomata are open it is inevitable that water vapour will diffuse out of the leaf down the diffusion gradient. At night, plants cannot photosynthesise so they close their stomata to conserve water. As light intensity increases, most plants open their stomata wider to obtain as much carbon dioxide as they can. This tends to increase the rate of transpiration.

Some desert plants open their stomata at night to take in carbon dioxide and then close them during the day to conserve water.

Measuring rates of transpiration

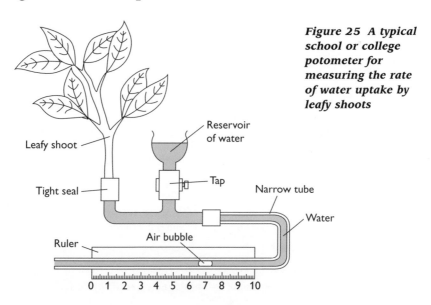

Figure 25 A typical school or college potometer for measuring the rate of water uptake by leafy shoots

You can use a potometer to measure the rate at which leafy shoots cut from plants absorb water. Potometers like that in Figure 25 measure rates of water uptake, *not* rates of transpiration. If they are put onto a balance, then they can measure *both* the rate of water uptake (by following the movement of the bubble of air) *and* the rate of transpiration (by measuring the loss in mass).

There are a number of precautions that should be taken if you are going to set up a potometer:
- The leafy shoot should be cut under water and placed into the potometer under water so that air does not enter the xylem vessels and block them.
- The leafy shoot should be left in the potometer for some time to adjust before any readings are taken.
- Conditions around the leafy shoot should remain constant while readings are taken. Repeat readings should be taken to ensure that the results are reliable.
- Conditions can be changed by altering the temperature, the humidity, the wind speed and the light intensity. While investigating one factor, such as temperature, the other factors should be kept constant. This is difficult to achieve in a school or college laboratory unless you have access to an environmental chamber that maintains constant conditions.

Readings taken from a potometer are given as distance travelled (by the air bubble) in a certain period of time. It is possible to calculate the volume of water that has been absorbed if you know the radius of the narrow tube. This may be 1 mm. If the bubble travels 20 mm in 15 minutes, the rate of water absorption is calculated as follows:

volume of a cylinder = $\pi r^2 h$ = 3.14 × 1 × 20 = 62.8 mm^3

rate of water uptake = $\dfrac{62.8}{15}$ = 4.2 mm^3 min^{-1} (to the nearest 0.1 mm^3 min^{-1})

Xerophytes

Xerophytes are plants that are adapted to living in places where there is a shortage of water. There are various adaptations to *reduce* water loss in the leaves of these plants, some of which are described in Table 9.

Feature	Adaptation for reduction of water loss
Leaf is permanently rolled or rolls up in dry conditions	Air is trapped inside the leaf; water vapour diffuses into the air, but is lost slowly to the atmosphere. The humid air is trapped and reduces the diffusion of water vapour from the stomata. Leaves that do this have their stomata facing inwards when the leaf is rolled.
Thick cuticle	Cuticle is made of waxy substances that waterproof the leaf
Leaf covered in hairs	Hairs trap a layer of still, humid air; this reduces diffusion of water vapour from the interior of the leaf
Stomata sunken in pits or grooves in the leaf	Still, humid air collects in the pits; this reduces the diffusion of water vapour through the stomata

Table 9 Some adaptations of xerophytic leaves

Translocation: source to sink

Plants make a great variety of organic compounds. Sucrose is the substance that they make for the transport of energy. Leaves make sucrose from the sugars they produce in photosynthesis. Sucrose travels from the mesophyll cells, where it is made, to the companion cells, which pump it across their membranes, and then it passes into sieve tube elements through plasmodesmata. This lowers the water potential inside the sieve tube elements so that water flows in from surrounding cells by osmosis. Hydrostatic pressure builds up inside the phloem sieve tubes and this forces the sugary solution from cell to cell through the sieve tubes and away from the leaves. Phloem sap flows from the leaves downwards to the roots and upwards to new leaves and to flowers and fruits. At these 'sinks', sucrose and other assimilates are removed from the sieve tubes and this lowers the hydrostatic pressure. This maintains a pressure gradient from source to sink. Phloem sap may move in opposite directions in adjacent sieve tubes, unlike the flow of water in xylem, which is always one way — upwards from roots to leaves. Mass flow in phloem is maintained by an active mechanism. There are several lines of evidence for this.

- The rate of flow is higher than can be accounted for by diffusion.
- Companion cells and sieve tube elements have mitochondria and use ATP to drive pumps to move sucrose. They achieve this by pumping H^+ out of the cell. H^+ diffuses back into the cell through a carrier protein, which also transports sucrose.

Links In the practical examination, you may be asked to look at microscope slides showing sections of root, stem or leaf. You may be asked to compare the leaf of a plant such as marram grass, which has adaptations for reducing water loss, with a leaf without these adaptations. A question that asks you to do this is testing you on topics from Module 2801 (plan diagram of a leaf) with information about xerophytes from Module 2803/01.

It is a good idea to revise the details about osmosis and water potential in Module 2801 for this section on transport in plants. You are expected to apply your understanding of these topics to the absorption of water from the soil by root hairs and the transfer of water through a plant. If asked about water, always use the terms 'osmosis' and 'water potential', and explain that water is moving '*down* a water potential gradient'. Details of plant cell structure from Module 2801 are also useful as you may have to label a drawing or diagram of phloem cells and you would need to recognise structures such as mitochondria, nuclei, endoplasmic reticulum and cell walls.

AS Biology

Experimental and investigative skills

The practical skills required for Modules 2803/02 (Coursework 1) and 2803/03 (Practical examination) are much the same. You should be given opportunities to develop these skills during the practical work in your AS course.

To help you with the four skills, we are going to look at how you might carry out a specific investigation on the effect of temperature on enzyme activity, which involves material from two learning outcomes in Module 2801. We have not included a complete piece of coursework as written by a candidate. Instead, there are suggestions about the steps that you should follow when undertaking coursework, illustrated with examples from this investigation. You can use the investigation as a training exercise to help you develop your practical skills and then apply the lessons you have learnt to your own coursework. For example, you will see that it is difficult to obtain precise results with the method described. This should make you think how to obtain precise results in your own investigation. You might like to refer to the Student Unit Guide for Unit 1 in this series, where the relevant theory is described (see pages 16–20 and 31–32).

There are references to the practical examination throughout the four sections. If you are taking this module, you need to plan, carry out experimental work, analyse results and evaluate procedures and data in the same way as those doing the coursework module. In addition, you should be able to make drawings and observations from the microscope (see pages 54–55).

Skill P: Planning

Step 1: Analyse (or 'unpick') the information you are given

Imagine you are given the following task:

Your task is to investigate the effect of temperature on the activity of amylase.

There are a number of ways in which this investigation could be tackled. You need to make some choices about how you are going to investigate this task. To do this, you should ask yourself a series of questions. These are outlined below. Some typical answers are given alongside, together with examiner comments, which are prefixed by the symbol 🖉.

Q What is amylase?
A It is an enzyme.

Q What does amylase do?
A It breaks down starch to sugar.

OCR Unit 3

> *e* Yes…but at AS we would expect you to have a more detailed knowledge than this. Try again.

A It breaks down starch to maltose by hydrolysis.

> *e* That's better!

Q What is the effect of temperature on enzymes?

A As the temperature increases, the enzyme works faster until it reaches the optimum temperature. After that, the enzyme is denatured and does not work.

> *e* In your planning you must show a good understanding of the vocabulary appropriate to AS. 'Denatured' is good. 'Works' — can we improve on that?

A The rate of the reaction increases as the temperature increases.

Q How do I measure the rate of reaction?

A From Biology Foundation, I know that I can find out how long it takes for the substrate to disappear or measure the product as it appears. I can use iodine solution to find out how long it takes for starch to disappear. I did this in class by taking out samples from a mixture of starch and amylase with a pipette at regular intervals and testing them with iodine solution in a spotting tile.

> *e* So, how do you know when the starch has been broken down?

A Because when the starch is still there in the sample, the iodine solution goes black (or blue). When the starch is all gone, the iodine solution doesn't change colour — it stays yellow. So I go on taking samples until the iodine solution stops going blue or black.

> *e* Very good. I think you understand this!

There are many other questions that you can ask yourself during this 'unpicking' of the task. Try using questions that begin with 'what', 'how', 'why', 'where', 'when' and so on. You should collect quantitative data in your investigation — ask yourself what outcome you can *measure*. In our example you can time how long it takes for all the starch to be broken down (see page 43). It is possible to improve this if you know how much starch has been broken down. You can then calculate the rate of breakdown, for example as milligrams of starch per minute.

It is not a good idea to collect qualitative data, such as the colours you see when using the Benedict's test for reducing sugars. Skill A involves carrying out calculations on your results. You cannot do this if the results you have taken comprise different colours.

Step 2: Identify the variables

It is a good idea to draw up a table to identify the variables in your investigation.

Type of variable	Explanation	Example for our investigation
Independent variable	This is the variable you are investigating and are going to change — sometimes called the input variable	Temperature
Dependent variable	This is what you will measure — sometimes called the output variable	Activity of the enzyme measured by time taken for starch to disappear
Control variables	Other variables that can influence the investigation and must be kept constant	• pH • Volumes of starch and amylase solutions • Concentrations of starch and amylase solutions

By keeping the control variables constant, you are designing a 'fair test'. You should make it clear in your strategy how you intend to keep these control variables constant. These details could be added to a table like the one above.

At AS you are expected to know that pH influences the activity of enzymes and you would be expected to control the pH of the reaction mixture. You can do this by using a buffer solution at a specific pH. In this investigation a buffer at pH 7.0 could be used. Buffer solutions can be prepared using tablets dissolved in distilled water or from various chemicals. You are not expected to know how to do this — you can ask your teacher and/or the laboratory technician for an appropriate buffer solution. If this is not possible, then you can make up all your solutions with distilled water and check that the pH of each reaction mixture is the same. You could use universal indicator paper or a pH meter.

Step 3: Plan a basic strategy

You could show this as a simple flow chart.

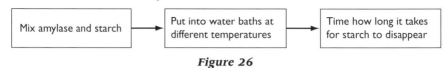

Figure 26

Decide how you will carry out this strategy in the laboratory. What concentrations of amylase and starch should you use? What volumes of the solutions should you use? What temperatures should you use? How do you find out how long it takes for starch to disappear? Keep asking questions like these about the practical work.

It is now time to organise some preliminary practical work to find answers to some or all of these questions. You should be given time in the laboratory to try out some preliminary work. This is sometimes called a pilot investigation. You should keep a careful record of what you do and the results that you obtain. If you are unsure about how to proceed, then you need to consult some textbooks and/or guides that give information about practical work. Your school or college biology department will have suitable books, as will your learning resources centre. You can also try your local library. You are likely to find suitable material on the web, but make sure you use reputable sites. Try Biozone:

http://www.biozone.co.uk

or Science and Plants for Schools (SAPS):

http://saps1.plantsci.cam.ac.uk

You might have done a class practical that will help with your planning. Keep details of the sources of information that you use and make sure that they are listed in the plan that you hand in.

Imagine we are carrying out a practical to follow the disappearance of starch. We will use 1% solutions of amylase and starch. As our preliminary experiment, we will try different volumes of starch solution and amylase solution at 30°C to see which give the best result. We are expecting the starch to go quite fast, as we know that 30°C is close to the optimum temperature for enzymes. We do not want the colour change to happen too quickly or we will not be able to measure it. We do not want it to be too slow, as the experiment will take too long. The results might be as follows:

Volume of 1% amylase solution/cm^3	Volume of 1% starch solution/cm^3	Results/observation
5	5	Starch all gone within 60 seconds
6	4	Starch all gone within 60 seconds
7	3	Starch all gone within 60 seconds
8	2	Starch nearly all gone within 60 seconds
9	1	Starch all gone within 120 seconds

You can see that the best volumes to use would be 9:1. You should draw up a table similar to this one to record your results and/or observations from your preliminary work.

Note that in this investigation, you have to take samples from the reaction mixtures and test them with iodine solution. This means having a pipette to take samples and putting them into a spotting tile.

When you have your preliminary results, you should be able to make some decisions about your strategy. Make sure that you have recorded these and that they are included in your plan.

As you work out your strategy, make a list of the apparatus and materials that you will need. There should be at least two ways in which you can carry out the investigation that has been set. This means you have to make some choices about the strategy, the apparatus to use and the quantities of materials. It should be clear in your plan why you have chosen your strategy and why you have selected certain key pieces of apparatus. You should justify the inclusion of certain pieces of apparatus by

AS Biology

referring to the precision of the results you will take. You can also justify the ways in which you will control variables. This means explaining *why* you have made these choices in your plan.

Step 4: Research some of the SKU that you need

You could find out why iodine solution is used to follow the disappearance of starch, or how to make your readings more precise by using a colorimeter. You can also research the background theory — see step 10 on page 44.

Step 5: Decide on the range for your independent (input) variable

In this investigation you should choose some cold temperatures at which there is little enzyme activity (0°C and 10°C), some near the optimum (20°C and 30°C), and some above the optimum to show that the enzyme is denatured at high temperatures (40°C and 50°C). The actual range for the independent variable will depend on the type of investigation you are carrying out. You might use your pilot investigation to help you decide on the range.

Step 6: Decide on the number of readings to take

At AS, you are expected to do at least five. Here, you could use five different temperatures, for example 10°C, 20°C, 30°C, 40°C and 50°C. However, you might want to find out whether there is any activity of amylase at 50°C. If so, then perhaps you should extend the range to 60°C or 70°C to check for complete denaturation of the enzyme. This would mean having seven different temperatures. It is a good idea to plan to have more than five readings across a suitable range.

Step 7: Ensure that the results you collect are reliable

Imagine carrying out our investigation and taking *one* result for each temperature. You will not know if this result is 'correct' or not. You can check this by repeating the reading. These are sometimes called 'check readings' or 'replicates'. Whatever you call them, put them in your plan and explain why they are there. You should repeat your investigation at least once; much better is to plan to repeat it twice. This means that your results table will look something like this.

Temperature/°C	Time taken for starch to disappear/seconds			
	1st expt	2nd expt	3rd expt	Mean
10				
20				

Remember that it is important to explain why you are planning to carry out replicates (check readings). You can do this in terms of the reliability of the evidence that you will collect.

Step 8: Decide on the level of precision with taking results

In your investigation, you are timing how long it takes for starch to disappear. You may decide to take samples from test-tubes at 1 minute intervals. If you have a positive result with iodine solution at 180 seconds and a negative result at the next sampling time of 240 seconds, then the starch must have finally disappeared sometime between these two sampling times. You cannot be more precise than this. An alternative way of doing the practical is to use a colorimeter to measure the absorbance of the sample when it is mixed with iodine solution. When there is a lot of starch present, the absorbance reading is high. When the starch has all been broken down, the absorbance reading is low. With this method, you must take samples with a pipette or a syringe from a reaction mixture in the same way as before. Although readings of the colour of the starch/iodine mixture can be measured more precisely than using your eye to judge, the timing is no more precise than taking samples with a pipette and testing in a spotting tile. You could improve the precision of the results by making sure you remove the same volume of reaction mixture and using the same volume of iodine solution each time. The volumes could be measured with a 1 cm^3 pipette.

You may be asked to investigate the effect of catalase, which is another enzyme mentioned in the specification for Module 2801. Here you can measure the activity of the enzyme by collecting oxygen, which is produced when catalase breaks down hydrogen peroxide:

$$2H_2O_2 \longrightarrow 2H_2O + O_2$$

You can use a measuring cylinder, a gas syringe or the barrel from a plastic syringe as a collecting vessel for the gas. Depending on the graduations on the collecting vessel, you may be able to measure the volume to the nearest 1.0 cm^3. This is a measure of the precision of your method.

Step 9: Carry out a risk assessment

You will need to carry out a risk assessment to check the safety aspects of your investigation. Your school or college may have a risk assessment form that you have to complete. If so, include this in your plan. If not, then you could set out a table like this.

Hazards	Precautions

Your school or college will probably have Hazcards, which are published by an organisation called CLEAPPS. You can use these for writing out your risk assessment. If you are going to implement your plan, then it is important that you follow the precautions you have given — and any others that your teacher insists on. *You must not start a piece of practical work until you are aware of the safety hazards and precautions to take.*

Step 10: Write a prediction

For nearly all investigations, it is possible to write a prediction stating what you expect to happen. For example:

> As temperature increases from 0°C to 30°C, the rate at which amylase breaks down starch increases. After 30°C, the rate decreases as amylase is denatured.

You should include a sketch graph if this makes your prediction clearer. Here you might include a graph like Figure 27.

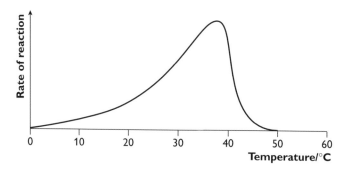

Figure 27 A student's prediction graph

The prediction may be suitable for some types of amylase used for AS practical investigations. In a preliminary investigation, you might have discovered that the optimum temperature for this enzyme is somewhere near 30°C. (Do not assume that the optimum temperature for all enzymes is 37°C; for example, some enzymes used for practicals in schools and colleges have much higher optimum temperatures.) It is then a good idea to justify this prediction with some SKU from your AS course. In this case, you could refer to kinetic energy, collisions between the enzyme and its substrate, active sites, glycosidic bonds in the starch molecule and hydrolysis of these bonds. You should not write a 'mini essay' on the relevant theory, but you should use your SKU to support what you have predicted.

Step 11: Write up your plan

The planning exercise in the practical examination has a 1000 word limit. There is no word limit for the coursework. However, you should always write concisely. Here are some suggestions to help you.

- Use tables wherever you can — for example, for variables, for results from the preliminary investigation, for showing how to make dilutions, and for showing how to collect results in the main investigation.
- Use numbered points when writing your method. This makes it easy for you to follow the steps in your plan when you are implementing it. It also helps others (your teacher, the moderator or the examiner) to follow your strategy.
- Use sketch graphs in your prediction and for showing how results are to be presented.

- Keep to the point by writing concisely. Show how you have used the SKU from your AS course. Do *not* write at length about the topic — start your plan by telling the reader briefly what your investigation is about and outline the main steps of your strategy.
- Give the results of your preliminary investigation (pilot study) and state clearly how these helped you to make decisions with your plan.
- State the sources of any information that you researched and used. You can list the textbooks and/or websites that you have used in a bibliography at the end, but you must indicate in the plan where you have *used* these sources. You may include direct quotes from these sources but these will not gain you any marks. State clearly how you have used the information from your sources in developing your strategy.
- Use plenty of sub-headings, so that your plan is well structured and easy to follow. Do not write dense prose.
- Write your own plan. Copying or adapting material from another source, such as from one of the websites that provides help with coursework, is plagiarism. Your teacher knows your work and will be suspicious if you do this. You are likely to be discovered and your work disqualified. If you use one or more of these websites, then include them as sources and make it clear how you used the information that you found. Note that material taken from websites may not be correct.

Skill I: Implementing

It is difficult to give you advice about this skill for your coursework since so much depends on how you carry out practical work in the laboratory. If you find practical work difficult, then here are some simple pieces of advice that might help you.

Following instructions

When you are given a set of instructions to follow, read them through to the end first. This will help you to anticipate what you are expected to do during the practical. It will also tell you what sort of results you are going to record. The instructions will be written so that you can follow them in sequence. Read the first instruction carefully and then carry it out. It is a good idea to put a tick by each instruction when you have completed it. Proceed carefully through the rest of the instructions, double-checking that you are doing the right thing.

Safety

You will be expected to work safely. You should be aware of the safety symbols that the laboratory staff may have put on glassware containing solutions that are hazardous. You should always wear a laboratory coat, and safety goggles when appropriate, and take special care when using sharp implements such as scalpels.

Working confidently

Take care over setting up your apparatus and measuring out volumes. Your teacher will assess this part of your work during the practical as it contributes to the mark for this skill in your coursework. You will not achieve the expected outcome from your practical work unless you take care over such things as measuring volumes of solutions.

Taking readings

Readings, measurements and observations should be taken carefully. The sorts of measurements that you might take are:
- length — with a ruler or a pair of callipers
- mass — with a balance
- volume — with a syringe, measuring cylinder or gas syringe
- optical density (absorbance or transmission) — using a colorimeter

In some cases, you may be recording the colour of a solution or estimating degrees of cloudiness. This might happen if you have used the Benedict's test to find a reducing sugar concentration or if you have used iodine solution to test for varying quantities of starch. If this is the case, then you need to be careful about writing down the colours. For degrees of cloudiness you can use a scale between 0 (no cloudiness, completely clear) and 10 (the most cloudy). However, you should realise that these are not quantitative results and they cannot be processed as described on page 47. It may be possible to make them quantitative by filtering, drying and finding the mass of the material that causes the cloudiness.

Recording

Your results and observations should be presented in a table. Figure 28 shows how you should do this.

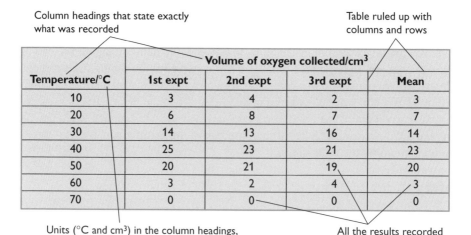

Column headings that state exactly what was recorded

Table ruled up with columns and rows

Temperature/°C	Volume of oxygen collected/cm³			
	1st expt	2nd expt	3rd expt	Mean
10	3	4	2	3
20	6	8	7	7
30	14	13	16	14
40	25	23	21	23
50	20	21	19	20
60	3	2	4	3
70	0	0	0	0

Units (°C and cm³) in the column headings, not in the middle of the table

All the results recorded

Figure 28

It is a good idea to give your table an informative title. You can start titles with 'Table to show...' If you have more than one table of results, then number them Table 1, Table 2 etc. This makes it easy to refer to them when writing your analysis.

Precision of measuring

The figures given in your table should indicate the degree of precision that you used. In the example above, we have measured to the nearest cm^3, so all the numbers are given as whole numbers without any decimal places. If you measured to the nearest 0.1 cm^3, then you would write 12.7, 2.3 etc. It would then *not* be appropriate to write 12. Instead you should write 12.0, indicating that you have measured to the same degree of precision as the other results.

As you carry out the practical aspects of your coursework, your teacher will assess how well you match the descriptors given in the specification for Skill I.

In the practical test, you gain Skill I marks for recording results in tables and for making observations. Some of the marks for Skill I may come from Question 2 if you are required to make drawings and observations from the microscope (see page 54). Your teacher does not assess your work during the test.

Skill A: Analysing

There are several steps that you can follow when analysing your results.

Step 1: Do some simple processing of your results

This will probably involve calculating the means for your check readings (see Figure 28). You may not have enough readings to be able to do this. It is quite acceptable to combine (or pool) your results with someone else's, or even with the rest of the class, so long as your teacher can identify which are your results. It is a good idea to include a table with your results and then a table with pooled results. You should then check for anomalous results, which is part of your evaluation — see page 51–52.

The mean values that you calculate should not be given to a greater degree of precision than the results that you have recorded. For example, if we measure some cores of potato as part of an investigation on osmosis and record these results:
    17 mm, 18 mm, 16 mm, 17 mm, 18 mm
then the mean is 17 mm, not 17.2 mm.

Step 2: Carry out some more detailed processing

The investigation that you are given should allow you to carry out further processing of your results. In the investigation of starch breakdown by amylase, we took results as lengths of time for starch to disappear. If these results are plotted on a graph, we would see something like Figure 29. This does not look like the graph showing the rate of reaction that we gave in our prediction (see Figure 27 on page 44). In this case, the further processing could involve calculating rates as $1/t$, where t is the time in seconds. You may wish to multiply the results of these calculations by 100, 1000 or 10 000 to give whole numbers that are easy to plot on a graph. The result will then have the same shape as the prediction graph.

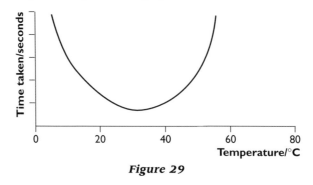

Figure 29

You can also calculate rates in terms of the quantity or volume produced per second. For example, the volume of gas collected when catalase breaks down hydrogen peroxide can be calculated as cm^3 per second — see Figure 30.

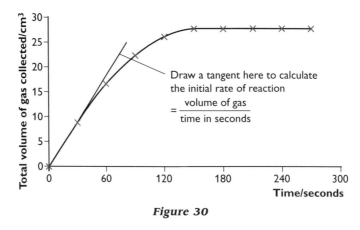

Figure 30

If you have collected several check readings for each value of the independent variable, you can calculate the standard deviation or the standard error. You may be confident about calculating these because you have been taught about them in another subject, such as maths or geography. You can use them to comment on the reliability of your results in your evaluation. When you carry out an experiment, your repeat readings are not always exactly the same. This is to be expected — you do not always carry out the procedure in *exactly the same way* each time and biological material varies, so there are bound to be differences between your repeat readings. Sometimes there may be so many sources of error that the repeat readings are quite different. This relates to reliability — how close or far apart are your repeat readings? Standard deviation or standard error is a useful way of *quantifying* the spread of repeat readings about the mean. If you do not know about standard deviation, you could just calculate the range of each set of check readings — 16–18 mm (= 2 mm) for the example given above — but this would not be enough for the detailed processing required. Calculating the initial rate of reaction, in the way shown in Figure 30, is an acceptable way to meet the requirement for detailed processing.

Statistical tests are not part of the AS biology course, so it is not expected that you will use any in your coursework. If you know of an appropriate statistical test (you may have come across one in geography or if you have done statistics at GCSE or in your AS maths course), then use it. But check with your teacher first, so that you use a test that is appropriate for the type of data you have collected.

Step 3: Draw a graph of your results

If you have investigated an independent variable such as temperature, pH or concentration, then you can draw a line graph of your results. If you investigate the activity of catalase from different sources, such as different animal and plant tissues, you will not be able to draw a line graph. Your independent variable (different sources) is not continuous and so you will have to draw a bar chart for your results. You may then find it difficult to match some of the descriptors for analysing and evaluating. You should carry out an investigation that allows you to draw line graphs since you can then comment on trends.

Figure 31 shows how a line graph should be drawn and presented.

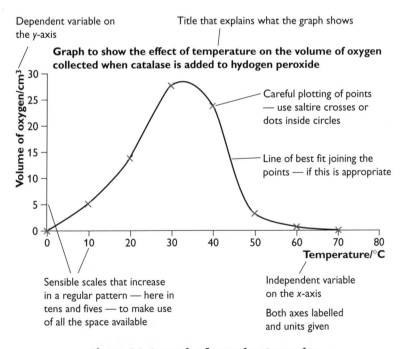

Figure 31 A graph of a student's results

When drawing your graph, you should consider whether the origin is a point on the graph and should be part of the line of best fit. In the example above, the student carried out a test at 0°C and found that no gas was collected. However, if this had not been done there would be no reason for starting the line at the origin. Sometimes it is obvious that the origin should be included. For example, if you are investigating the effect of the substrate concentration on the rate of an enzyme reaction, then the origin (0,0) must be a point, since there will be no reaction if there is no substrate present!

You may wish to use a software package to draw your graphs. It is often difficult to get acceptable graphs this way, especially if you wish to use them to carry out further processing or to read some figures from the graph. You are advised to draw your main graph by hand on graph paper. If you are going to include a series of graphs, then at least one should be hand drawn.

Step 4: Making use of the graph

Sometimes you can use your graph to do some further processing of your results. For example, plotting a time course graph such as that in Figure 30 allows you to find the initial rate of the reaction by drawing a tangent to the line and calculating the gradient. In an osmosis experiment, you may be able to find the water potential of a plant tissue by using the intercept on the graph, as shown in Figure 32.

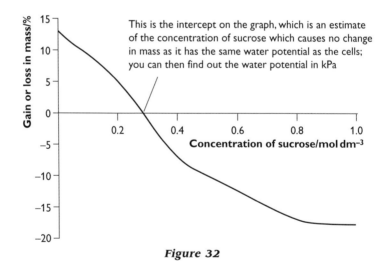

Figure 32

Step 5: Look for a trend (or trends) in your graph (or graphs)

Describe the trend in words, using results from the graph (or from your table) to illustrate it. You must now show relevant scientific knowledge and understanding (SKU) to explain the results that you have obtained. You may have written all about this in your plan, in which case you should recap on what you wrote and refer back to it. It

is good to keep the SKU to a minimum in the planning and keep most of it for the analysis. As you write your analysis, take care to:
- write accurately, paying attention to spelling, punctuation and grammar
- use your results to illustrate trends
- use specialist terms correctly
- use SKU from the AS course to underpin the conclusions that you draw
- write concisely and coherently

In the practical test, you will be asked to draw graphs of your own results or of data that are printed in the paper. You will also be asked to describe your results and possibly compare them with given data. There will be questions that ask you to explain the results and draw conclusions, relying on your SKU from the AS course.

Skill E: Evaluating

There are two distinct strands to this skill:
- evaluating the method that you used
- evaluating the data that you collected

It is a good idea to write your evaluation in two separate sections. The first section should deal with the overall strategy you designed (or were given), the way you set up the apparatus and materials and the way in which you collected results. The second section should evaluate the data that you collected. You can end by commenting on how confident you are about your conclusions, bearing in mind the criticisms that you have made of the procedure and the results.

You should think about these sorts of questions when you are preparing your evaluation.
- What are the limitations of the method?
- What impact might these limitations have on the results?
- What improvements can you suggest for the limitations you identify?
- Are there any anomalous results?
- What may have been responsible for these anomalous results?
- How reliable are the results?
- How certain can you be of your conclusions?

Make a list of the ideas that you have. The first step is to identify any anomalous results. In fact, you should have done this earlier, before processing your results. An anomalous result is one that does not fit the trend. You are likely to find two sorts of anomalous result:
- a check reading that does not agree with the others, for example with readings of 11, 15, 13, 14, 3, you should not include the '3' in the mean that you calculate; this type of anomaly can be ringed in your table of results
- a result that does not fit the trend when plotted on a graph — see Figure 33.

AS Biology

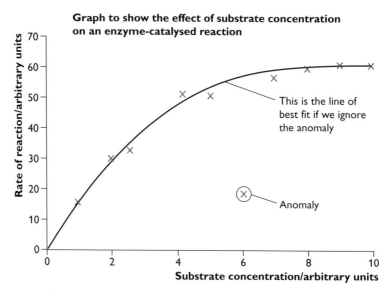

Figure 33 *Obviously the point at 6 is an anomaly and should not be included in the line of best fit*

If you have checked and there are no anomalous results, then you should say so. If you find some, then you should try to explain the reasons for these anomalies. As you carry out the investigation, you should think critically about what you are doing, try to recognise anomalous results and look for something that you have done that may explain them. They may be quite obvious: if collecting gas, you may have a leak in your apparatus; you may have measured out the volumes incorrectly or labelled a test-tube incorrectly.

Make a list of the limitations of your method. These might have to do with the overall design of the strategy, with the way you set up each experiment or with taking your results. Excuses such as 'there was not enough time' are not appropriate here and do not gain any credit! Decide which of these limitations are the most important. You could rank them in order of importance. There is usually some aspect of your overall strategy that is the major error. For example, you may have added enzyme to substrate and not been able to measure the product right at the beginning of the reaction. This is often the case when you are adding catalase to hydrogen peroxide.

Suggest ways in which each limitation you have identified could be addressed and explain how the improvement will reduce the errors. You can do all this with a table:

Limitation	Way to improve the limitation	Reason for the improvement

You can put the limitations in order of decreasing importance and should always identify the most significant one(s). Do not list a large number of limitations: many are likely to be trivial.

Comment on the accuracy and reliability of your results. Accuracy and reliability are *not* the same. There are two aspects of accuracy:
- how close the results are to 'true' values
- the degree of precision of your results

It is sometimes difficult to know how close your results are to the 'true' values since you cannot look these up in a textbook or calculate them using a formula. However, you should have a predicted trend. You can draw this on your graph of results and compare it with your line of best fit. For example, you may have predicted a plateau in your graph, but your own results do not show this effect.

It is possible to calculate the percentage error for your results. Imagine that you have collected a gas and measured the volume with a syringe that has graduations every $1\,cm^3$. If you have measured $5\,cm^3$ of gas with your syringe, then you can be certain that you have more than $4.5\,cm^3$ but less than $5.5\,cm^3$. Your error is $+/-\,0.5\,cm^3$ in $5\,cm^3$. This makes the percentage error:

$$\frac{0.5}{5} \times 100 = 10\%$$

If the volume of gas collected was $10\,cm^3$, then the percentage error would be 5%.

You can also comment on the reliability of your results by stating how close the individual check readings are to each other or by using the calculations of standard deviation or standard error to quantify the reliability of your data. If your check readings are all close together, you can make a positive comment and say that you consider them to be reliable.

If your results are not reliable and you think they were taken with a low degree of precision, then you probably cannot be confident in your conclusions. This means that your conclusions are unlikely to be valid. You can quantify this by saying that the results are likely to be an underestimate or an overestimate of the 'true' values. You may not have enough data to be able to make a satisfactory conclusion because you did not take enough readings across your range for the independent variable. For example, it may not be possible to state the optimum temperature for an enzyme. It may lie somewhere between 30°C and 40°C. You should finish your evaluation by commenting on such limits of your conclusion(s).

In the practical test, you will be asked to evaluate the practical you have carried out and the results you have obtained and/or been given. You can follow the same procedure as outlined above and comment on such things as anomalous results, precision, reliability and improvements.

Microscopy in the practical test

One of the questions in the practical test will involve work with the microscope. This question should take you about 30 minutes. When you start the question, read it all through carefully so you know what is expected of you. Look at the slide with your naked eye first before putting it on the microscope. This may help you position it on the stage of the microscope. You will often be told what is on the microscope slide, so you will not have to identify it. You will also be told if it is something that you are unlikely to have seen before. Don't panic — it will be similar to something that you should have seen. You are likely to be given a drawing or a photograph to help with some aspect of the question.

When you make a drawing, follow these simple rules:
- Low-power plans (e.g. of a leaf) only show the outlines of the tissues — they do not include drawings of cells.
- High-power drawings should show a small number of cells (three or four maximum) and they should be drawn a reasonable size so you can show any detail inside them.
- Use a sharp HB pencil and draw single lines.
- Do not use any shading.

Add labels and annotations (notes) to your drawing *only* if you are asked for these in the question. Use a ruler to draw straight lines from the drawing to your labels and notes.

Figure 34 is a photograph of a leaf. Figure 35 is a low-power plan drawn from a slide and Figure 36 is a drawing of three of the palisade mesophyll cells as seen under high power. The drawings and annotations are similar to those that might be made by a candidate.

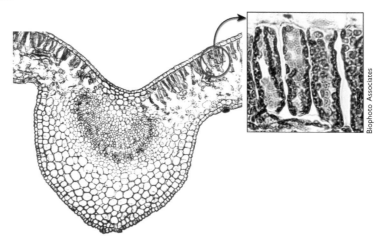

Figure 34 This is a cross-section of a leaf of privet. The inset shows some palisade mesophyll cells as seen with the high power of the microscope.

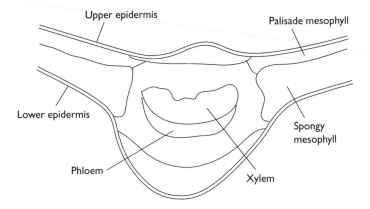

Figure 35 A tissue map drawn from a slide of a leaf similar to that shown in Figure 34. An examiner would award marks for showing the tissues in the leaf without including any cells. There would also be marks for labelling the tissues.

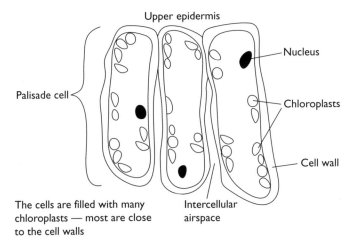

Figure 36 A high-power drawing of three palisade mesophyll cells from Figure 34. An examiner would credit the shape of these cells (longer than wider), showing the cell walls, the nuclei and the chloroplasts, and for the labelling and annotations.

Questions & Answers

AS Biology

A total of 45 marks is available in the module test. To give you an idea of the range of questions that may be set on the three sections in this module, the questions here total 60 marks.

As you read through this section, you will find that Candidate A gains full marks for all the questions. This is so that you can see what high-grade answers look like. Remember that the minimum for a grade A is about 80% of the maximum mark for the paper (36 out of 45 for the module test). Candidate B makes a lot of mistakes — these are ones that examiners encounter quite frequently. I will tell you how many marks Candidate B gets for each question. The minimum for a grade E is about 40% of the maximum mark (18 marks out of 45 in the module test). Candidate B's overall mark for these questions is just under 40%, which means that he/she will not have quite enough marks to gain a grade E.

Before you work your way through these questions and answers, it might be a good idea to reread the note about units on page 10.

Examiner's comments

Candidates' answers are followed by examiner's comments. These are preceded by the icon 🅔 and indicate where credit is due. In the weaker answers they also point out areas for improvement, specific problems and common errors, such as lack of clarity, weak or non-existent development, irrelevance, misinterpretation of the question and mistaken meanings of terms.

Question 1

Blood, tissue fluid and lymph

(a) The table gives differences between the composition of blood, tissue fluid and lymph. Complete the table. (4 marks)

Feature	Blood	Tissue fluid	Lymph
Red blood cells	Present		
White blood cells			Very few
Plasma proteins		Very few	
Haemoglobin		Absent	

(b) Red blood cells are involved in the transport of gases. Statements A to C describe some of the features of red blood cells.
(A) Red blood cells are biconcave discs.
(B) Red blood cells have no nucleus, mitochondria or endoplasmic reticulum.
(C) The cytoplasm of red blood cells contains molecules of the enzyme carbonic anhydrase.
Explain how each of these features (A–C) helps red blood cells to transport gases. (6 marks)

(c) Figure 1 is a drawing of a white blood cell and some red blood cells viewed with a light microscope.

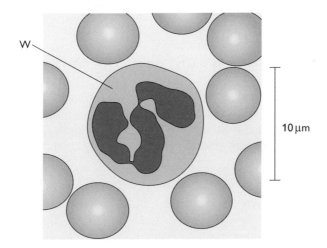

Figure 1

(i) Calculate the magnification of the drawing. Show your working and express your answer to the nearest whole number. (2 marks)
(ii) State the name of the cell labelled **W** and give a reason for your answer. (2 marks)

Total: 14 marks

AS Biology

question

Candidates' answers to Question 1

Candidate A
(a)

Feature	Blood	Tissue fluid	Lymph
Red blood cells	Present	Absent	Absent
White blood cells	Present	Some present	Very few
Plasma proteins	Present	Very few	Very few
Haemoglobin	Present	Absent	Absent

Candidate B
(a)

Feature	Blood	Tissue fluid	Lymph
Red blood cells	Present	✗	✗
White blood cells	✓	✓	Very few
Plasma proteins	✓	Very few	Some
Haemoglobin	✓	Absent	✗

👉 Candidate B scores 3 marks. The question did not ask for ticks and crosses. The table in the question has words such as 'present', 'absent' and 'very few', and this is a hint that candidates should complete the table in the same way. Here, the examiner has not accepted a tick for 'some present' for white blood cells in tissue fluid, but the ticks and crosses have been accepted, reluctantly, for the other answers.

Candidate A
(b) **(A)** This gives red blood cells a large surface area for diffusion of oxygen and carbon dioxide.
(B) Without these organelles there is more space for haemoglobin and therefore more oxygen can be carried.
(C) Carbonic anhydrase is an enzyme that causes carbon dioxide to react with water to produce hydrogencarbonate ions. This helps the blood to carry a lot of carbon dioxide.

Candidate B
(b) **(A)** Red blood cells have a biconcave disc shape to give them a large surface area to volume ratio.
(B) Red blood cells cannot divide or make their own proteins.
(C) Carbonic anhydrase helps in the transport of carbon dioxide.

👉 For (A), Candidate B has not explained how the large surface area is useful in terms of transporting oxygen — only part of the answer has been given and so gains 1 mark. The answer to (B) is not relevant — these are other consequences of the points made in the question. (C) is correct, but does not gain a mark because it does not provide enough detail. The candidate should have explained *how* the enzyme helps with the transport of carbon dioxide. When you are asked about 'adaptations', make sure that you state the feature, for example large surface area, and link it to a function, for example diffusion of oxygen.

Candidate A

(c) (i) Actual size is 10 μm. The length in Figure 1 is 30 mm, which is 30 000 μm.

$$\text{magnification} = \frac{\text{length of image}}{\text{actual length}} = \frac{30\,000}{10}$$
$$= 3000$$

Candidate B

(c) (i) The length in Figure 1 is 3 cm.

$$\frac{3000}{10} = 300$$

☒ Candidate B has measured the length of the scale bar in centimetres and then multiplied by 1000, thinking that there are 1000 micrometres in a centimetre! There are 10 000 micrometres in a centimetre. When doing these questions, it is much better to measure in millimetres and *not* in centimetres.

Candidate A

(c) (ii) Neutrophil; it has a lobed nucleus.

Candidate B

(c) (ii) Phagocyte; it has a nucleus, unlike the surrounding red blood cells.

☒ Candidate B gains a mark for the identification but other types of blood cell, for example lymphocyte, have nuclei, so the reason does not gain a mark.

☒ **Candidate B scores 5 marks out of 14 for this question.**

AS Biology

Transport of oxygen

Llamas are mammals that are adapted to live at high altitude in the Andes in South America. They are often found at altitudes higher than 5000 metres above sea level. Figure 1 shows the oxygen haemoglobin dissociation curves for a human at sea level and a llama that lives at high altitude.

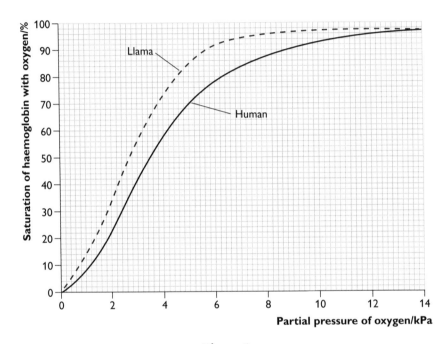

Figure 1

(a) At 5000 metres, the partial pressure of oxygen in the lungs is about 6 kPa. Using Figure 1, state the percentage saturation of haemoglobin with oxygen at a partial pressure of oxygen of 6 kPa for both humans and llamas. (2 marks)

(b) Explain the advantage, in terms of oxygen transport, of the dissociation curve for the llama being to the left of that for humans. (2 marks)

(c) The dissociation curves shown in Figure 1 are described as sigmoid or S-shaped. Using the information in Figure 1, explain the advantage of this shape in terms of oxygen supply to the tissues. (3 marks)

(d) When humans move to high altitude, the number of red blood cells increases. Explain how this helps humans to survive at high altitudes. (2 marks)

(e) State *three* ways in which the blood transports carbon dioxide. (3 marks)

Total: 12 marks

OCR Unit 3

Candidates' answers to Question 2

Candidate A

(a) Llama 92%, human 78%

Candidate B

(a) Llama 92%, human 79%

> 🖉 It should be easy to read the numbers from the graph, but if you find this difficult then use a ruler and a pencil. Rule a vertical line upwards from 6 kPa and then rule two horizontal lines from the points where your line hits the curves to the vertical axis. Look carefully at the scale, work out the answer and then check it. Here, one 'little square' = 2% on the vertical axis. Examiners often allow a range of answers for this type of question. Here they would allow 91–93% and 77–79%. Candidate B therefore gains 2 marks.

Candidate A

(b) There is very little oxygen in the atmosphere at 5000 metres. The llama's haemoglobin is nearly fully saturated with oxygen at the partial pressure at that altitude. This means that the haemoglobin carries enough oxygen for respiration.

Candidate B

(b) The llama haemoglobin carries more oxygen than human haemoglobin.

> 🖉 Candidate A has used the information in the question well and has picked up the cue from part (a). The answer to part (a) shows that at the pO_2 in the lungs, the llama's haemoglobin has a greater affinity for oxygen. This means that at this partial pressure, human haemoglobin carries less oxygen. Candidate B does not explain why the greater affinity of the llama haemoglobin is an advantage for living at altitude, but gains 1 mark for the statement given.

Candidate A

(c) Tissues have a low concentration of oxygen. When there are high rates of respiration, such as during exercise, this concentration decreases. A small decrease in the concentration of oxygen leads to a big decrease in the percentage saturation of haemoglobin. For example, from 4 to 2 kPa the haemoglobin gives up about 35% of its oxygen — the steepest bit of the curve. The oxygen then diffuses into the tissues.

Candidate B

(c) Haemoglobin combines with oxygen to form oxyhaemoglobin. As blood flows through capillaries in the tissues, the oxyhaemoglobin gives out its oxygen, which diffuses into the tissues.

> 🖉 Candidate A has used the information in Figure 1, as some figures from the graph have been given, but Candidate B has just stated some facts about oxygen transport, without using the graph. However, Candidate B gains 2 marks, for explaining that oxyhaemoglobin 'gives out oxygen' and for 'diffuses into the tissues'. Candidate B should have used the term 'dissociates' rather than 'gives out'.

Candidate A

(d) Haemoglobin is not as saturated at high altitude as at sea level. Having more red blood cells makes up for this because it means there will be more haemoglobin to carry some oxygen, although none of it will be 90%+ saturated.

Candidate B

(d) With more red blood cells, more oxygen can be transported as oxygen combines with haemoglobin.

> Candidate A has given a good explanation here. Candidate B has not used the idea of 'saturation' and again has not realised that the graph helps get to the answer. However, Candidate B gets 1 mark for stating the idea that more oxygen can be transported.

Candidate A

(e) • Combined with haemoglobin — carbamino-haemoglobin
   • As hydrogen carbonate ions in the plasma
   • CO_2 in solution in the plasma

Candidate B

(e) Carbon dioxide diffuses into the blood and combines with haemoglobin. Some of it reacts with water to form carbonic acid.

> Candidate A states three good answers. Candidate B has put in some unnecessary material — 'carbon dioxide diffuses into the blood' — but has stated that it combines with haemoglobin. Examiners would expect candidates to say that carbon dioxide forms the hydrogencarbonate ion. Candidate B gains 1 mark here.

> **Candidate B scores 7 marks out of 12.**

Heart structure and the cardiac cycle

Figure 1 shows an external view of a mammalian heart.

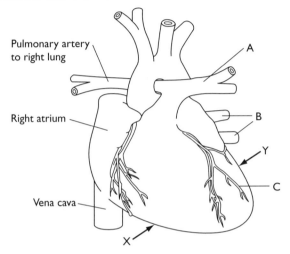

Figure 1

(a) Name the blood vessels **A** to **C**. (3 marks)
(b) State the position of the sino-atrial node in the heart. (1 mark)

Figure 2 is a life-size drawing of a cross-section of the heart of a young mammal in the region indicated by the arrows labelled **X** and **Y** in Figure 1.

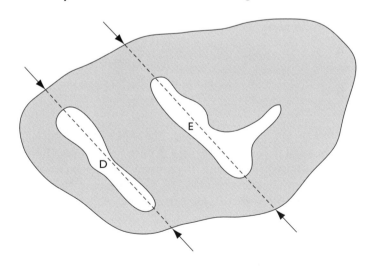

Figure 2

65

AS Biology

question 3

(c) Complete the table by measuring the total thickness of the wall of chamber E along the dotted lines indicated in Figure 2. (1 mark)

Chamber	Total thickness of wall/mm
D	14
E	

(d) Explain why the wall of chamber E is much thicker than the wall of chamber D. (2 marks)

(e) Describe the events that occur during the cardiac cycle. In this question, **1 mark is available for the quality of written communication.** (8 marks)

Total: 15 marks

■ ■ ■

Candidates' answers to Question 3

Candidate A
(a) A: Pulmonary artery
   B: Pulmonary vein
   C: Coronary artery

Candidate B
(a) A: Aorta
   B: Pulmonary artery
   C: Cardiac artery

> Candidate B has confused the aorta and pulmonary artery. Check with page 26 if you cannot see why Candidate B's answers are wrong. The heart is made of *cardiac* muscle; the blood vessels supplying this muscle are *coronary* arteries. Candidate B scores no marks here.

Candidate A
(b) Right atrium

Candidate B
(b) Atrium

> Candidate B has not been specific enough. The SAN is situated in the *right* atrium. You may be asked to put a label line on a drawing such as Figure 1 on page 65 to indicate the position of a structure, in which case the question would not have an answer line. Many candidates fail to answer these questions. This may be because they look for dotted answer lines for each question.

Candidate A
(c) 24 mm

Candidate B
(c) 60 mm

OCR Unit 3

👉 Candidate B has measured the whole thickness of the heart, not just the wall. The examiner has given a clue by including the wall thickness for chamber D. When you are asked to complete a table, always look at the rest of the figures in the table for clues about what to do.

Candidate A
(d) The left ventricle (E) is thicker because it has more muscle. When it contracts it raises the blood pressure higher than that in the right ventricle. This is because it has to force the blood around the whole body.

Candidate B
(d) The right ventricle (D) is thinner as it only forces blood to the lungs.

👉 Candidate A has answered the question. Candidate B has answered a different question, but would gain 1 mark for giving the 'reverse argument' for the fact that blood only goes as far as the lungs. The second mark is not awarded as there is no reference to pressure. It is *not* a good idea to answer a question in the way that Candidate B has done. The question asks about chamber E.

Candidate A
(e) As the cardiac cycle is a continuous process, we can start anywhere in the cycle. First, blood flows from the pulmonary veins into the left atrium. Valves prevent backflow of blood into veins. The atria contract (left and right together) — this is atrial systole. When the pressure is greater in the atria than the ventricles, the blood pressure pushes open the atrio-ventricular valves, forcing blood into the ventricles of the heart.

About 0.1 s later, the ventricles contract — this is ventricular systole. When the pressure is greater in the left ventricle than the atrium, the atrio-ventricular valve closes, and the high pressure of the blood (about 14 kPa) forces open the semi-lunar valves; blood flows from the left ventricle into the aorta.

Then the ventricles relax — this is ventricular diastole. Here the semi-lunar valves close when there is a higher pressure in the arteries than the ventricles. Also, blood flows from the pulmonary veins into the left atrium, again under low pressure, and the blood supply slowly trickles into the ventricle from the atrium. The process starts again.

Semi-lunar valves prevent backflow of blood into the ventricles. The semi-lunar valves close when blood fills their cusps.

Candidate B
(e) The blood flows into the heart through the vena cava. It enters the right atrium and is deoxygenated blood. The right atrium pushes the blood into the right ventricle from where it goes to the lungs. After it is oxygenated in the lungs it returns to the heart along the veins into the left atrium. Then it enters the left ventricle which pumps blood into the aorta and around the rest of the body. Valves prevent backflow in the heart.

The heart is made of cardiac muscle and has a sino-atrial node that controls when and how it beats.

question 3

AS Biology

🖉 Candidate A has given a concise and detailed answer and easily scores full marks and gains the quality mark. Candidate B has made the mistake of describing the pathway of blood through the heart, but gains 2 marks for stating that blood flows from veins into the atria and that valves in the heart prevent backflow. However, Candidate B has not written much and has not used the appropriate specialist terms, so does not gain the quality mark. When asked to write about the cardiac cycle, you should describe the events that happen in the heart as a whole. For example, state that the ventricles contract at the same time to force blood from the left ventricle into the aorta and from the right ventricle into the pulmonary arteries. However, it is possible to write about the events that occur in just one side of the heart and still gain the marks. Candidate A has done this by referring to the left side of the heart. But make this clear to the examiner. Remember to get the right sequence of events (see pages 28–30).

🖉 **Candidate B scores only 3 marks out of 15 for this question.**

Transpiration and the transport of water

(a) Define the term *transpiration*. (2 marks)

Figure 1 shows the pathway taken by water molecules from the xylem in a leaf to the atmosphere.

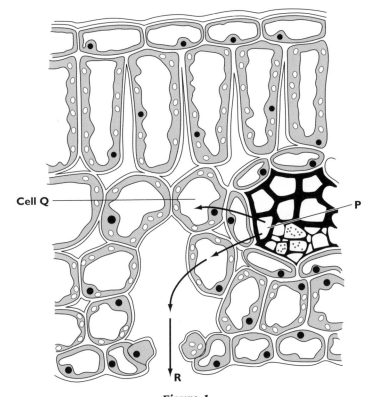

Figure 1

(b) Explain how water is transported in the xylem. (2 marks)
(c) Explain, using the term water potential, how water travels from the xylem vessel at **P** to cell **Q**. (3 marks)
(d) State what is shown by the arrow **R**. (1 mark)
(e) Explain how *two named environmental* factors influence the rate of transpiration from leaves. (4 marks)
(f) Describe *one* way in which the leaves of xerophytes reduce the rate of transpiration. (1 mark)

Total: 13 marks

AS Biology

Candidates' answers to Question 4

Candidate A
(a) Transpiration is the loss of water vapour from the leaves and stem of a plant.

Candidate B
(a) It is when water is lost from plants.

> 🖉 Candidate A gains full marks by referring to water *vapour* and to the parts of the plant where it is lost. Candidate B does not gain either mark as the word *vapour* is missing and so is any indication of the parts of the plant involved. It would also be a good idea to state that water vapour is lost by diffusion through stomata in leaves.

Candidate A
(b) Water molecules evaporate from the leaf and this pulls water molecules along the xylem because they are 'stuck' to each other by hydrogen bonds. This is cohesion-tension.

Candidate B
(b) Water is absorbed by root hairs by osmosis. Water enters the xylem in the root and passes through vascular bundles to the leaves where it evaporates.

> 🖉 Candidate B has explained how water enters the root hairs, which is not asked by the question. The answer then describes the *pathway* taken by water and what happens to it once it reaches the leaves. The candidate has not answered the question and so gains no marks.

Candidate A
(c) The cells in the leaf have a lower water potential than the contents of the xylem because cells like Q are probably losing water to the atmosphere. They also have sugars and salts inside to give them a low water potential. Water diffuses by osmosis down a water potential gradient into cell Q.

Candidate B
(c) Water moves by osmosis along a water potential gradient.

> 🖉 Candidate A has stated the direction of the water potential gradient and has explained why Q has a *lower* water potential than the xylem vessel. Candidate B gains a mark for stating that the movement is by osmosis, but gains no credit for 'along' a water potential gradient — it must be 'down' or from a higher to a lower water potential.

Candidate A
(d) R — diffusion of water vapour

Candidate B
(d) R — oxygen leaving the leaf

> 🖉 Although arrow R does show the direction taken by oxygen leaving the leaf *when it is illuminated*, we are not told in the question that the leaf is in the light. The

introduction to the question says 'the pathway taken by water', so Candidate B does not gain a mark here.

Candidate A

(e) • Light — stomata open in the light and so water vapour can escape.
 • Humidity — when the air is very humid there is no gradient between the saturated air inside the leaf and the air outside, so water vapour does not diffuse out.

Candidate B

(e) • Temperature — when the temperature is high, the water molecules have more kinetic energy and more of them evaporate.
 • Wind speed — on a windy day the water vapour molecules get blown away as soon as they come through the stomata. This means even more come out and the transpiration rate is high.

 ✎ Both candidates gain all 4 marks here. They identify the factors first and then explain their effects.

Candidate A

(f) Stomata sunken into pits

Candidate B

(f) Deep and extensive roots that spread out just under the surface of the soil

 ✎ Candidate B has not read the question. The learning outcome in the specification is about the adaptations of *leaves* of xerophytes and this is what the question asks about. Although Candidate B gives a correct statement about xerophytes, he/she gains no marks for this answer.

 ✎ **Overall, Candidate B gains 5 marks out of 13 for Question 4.**

AS Biology

Translocation

(a) Complete the passage about translocation in the phloem by using the most appropriate word (or words).
Assimilates, such as amino acids and _____ are transported from leaves, also know as 'sources', to areas known as _____ such as roots and fruits. Assimilates are loaded into sieve tube elements by _____ cells. The water potential inside the sieve tubes decreases so that water enters by _____, so increasing the hydrostatic pressure. This forces phloem sap through the sieve tubes. The sieve tube elements have cross walls perforated by _____. (5 marks)

(b) It is thought that sap moves through the phloem by mass flow. State one piece of evidence that suggests that the movement of sap in phloem is by mass flow. (1 mark)

Total: 6 marks

Candidates' answers to Question 5

Candidate A
(a) Sucrose; sinks; companion; osmosis; sieve pores

Candidate B
(a) Sugars; sinks; companion; diffusion; sieve plates

> Candidate B needs to be more specific. The main assimilate transported in the phloem is sucrose. Sugars could include glucose, which is wrong. Similarly, water enters the sieve tube elements by osmosis from surrounding cells. Diffusion is not quite specific enough here. Sieve plates are the cross walls. The question asked for the perforations (holes), which are called sieve pores. Candidate B scores 2 marks.

Candidate A
(b) Phloem sap moves faster than the movement of substances by diffusion. Diffusion occurs very slowly. The contents of one sieve tube move in the same direction by mass flow.

Candidate B
(b) Phloem sieve tubes are living and have sieve plates to help the movement of the phloem sap.

> Candidate A has given two pieces of evidence. Both are correct, but if the first one had been incorrect, then no marks would have been awarded. Candidate B has not given enough information. Phloem sieve tubes contain cytoplasm and are alive, but this does not give any evidence in support of mass flow.

> Candidate B gains 2 marks out of 6 for Question 5.

OCR Unit 3

☝ Overall, Candidate B scores 22 marks, which is just under 40% of the total marks available for these questions. This is unlikely to be enough to gain a grade E. Notice that, with a little more care, Candidate B could have gained more marks. Marks were lost for a number of different reasons:
- Some answers are not developed in full, for example Q.2(b), Q.3(e).
- Appropriate terms are not used, for example Q.2(e), Q.3(a).
- Calculations are not carried out correctly, for example Q.1(c)(i).
- Instructions are not followed carefully, for example Q.1(a), Q.3(c and d), Q.4(b).
- Data provided have not been used in answers, for example Q.2(c).
- Answers are not specific enough, for example Q.3(b), Q.5(a).

These are all examples of poor examination technique. With practice, you can avoid making these errors by reading the examination paper carefully and checking over your answers. With more care, Candidate B could easily gain enough marks to reach grade **C** or better.